Agnes E. Rupley, DVM, Dipl. ABVP–Avian
CONSULTING EDITOR

# VETERINARY CLINICS OF NORTH AMERICA

Exotic Animal Practice

Cytology

GUEST EDITOR
Michael M. Garner, DVM, Dipl. ACVP

January 2007 • Volume 10 • Number 1

**SAUNDERS**
An Imprint of Elsevier, Inc.
PHILADELPHIA   LONDON   TORONTO   MONTREAL   SYDNEY   TOKYO

**W.B. SAUNDERS COMPANY**
A Division of Elsevier Inc.

Elsevier, Inc., 1600 John F. Kennedy Blvd., Suite 1800, Philadelphia, PA 19103-2899

http://www.vetexotic.theclinics.com

**VETERINARY CLINICS OF NORTH AMERICA:**  Volume 10, Number 1
**EXOTIC ANIMAL PRACTICE**  ISSN 1094-9194
January 2007  ISBN-13: 978-1-4160-4379-9
Editor: John Vassallo; j.vassallo@elsevier.com  ISBN-10: 1-4160-4379-9

Copyright © 2007 by Elsevier Inc. All rights reserved. No part of this publication may be reproduced or transmitted in any form or by any means, electronic or mechanical, including photocopy, recording, or any information retrieval system, without written permission from the publisher.

Single photocopies of single articles may be made for personal use as allowed by national copyright laws. Permission of the publisher and payment of a fee is required for all other photocopying, including multiple or systematic copying, copying for advertising or promotional purposes, resale, and all forms of document delivery. Special rates are available for educational institutions that wish to make photocopies for nonprofit educational classroom use. Permissions may be sought directly from Elsevier's Rights Department in Philadelphia, PA, USA: phone: (+1) 215 239 3804, fax: (+1) 215 239 3805, e-mail: healthpermissions@elsevier.com. Requests may also be completed on-line via the Elsevier homepage (http://www.elsevier.com/locate/permissions). In the USA, users may clear permissions and make payments through the Copyright Clearance Center, Inc, 222 Rosewood Drive, Danvers, MA 01923, USA; phone: (978) 750-8400; fax: (978) 750-4744, and in the UK through the Copyright Licensing Agency Rapid Clearance Service (CLARCS), 90 Tottenham Court Road, London W1P 0LP, UK; phone: (+44) 171 436 5931; fax: (+44) 171 436 3986. Others countries may have a local reprographic rights agency for payments.

Reprints. For copies of 100 or more of articles in this publication, please contact the commercial Reprints Department, Elsevier Inc., 360 Park Avenue South, New York, New York 10010-1710. Tel: (212) 633-3813 Fax: (212) 633-1935, e-mail: reprints@elsevier.com.

The ideas and opinions expressed in *Veterinary Clinics of North America: Exotic Animal Practice* do not necessarily reflect those of the Publisher. The Publisher does not assume any responsibility for any injury and/or damage to persons or property arising out of or related to any use of the material contained in this periodical. The reader is advised to check the appropriate medical literature and the product information currently provided by the manufacturer of each drug to be administered to verify the dosage, the method and duration of administration, or contraindications. It is the responsibility of the treating physician or other health care professional, relying on independent experience and knowledge of the patient, to determine drug dosages and the best treatment for the patient. Mention of any product in this issue should not be construed as endorsement by the contributors, editors, or the Publisher of the product or manufacturers' claims.

*Veterinary Clinics of North America: Exotic Animal Practice* (ISSN 1094-9194) is published in January, May, and September by Elsevier, Inc.; Business and Editorial offices: 1600 John F. Kennedy Blvd., Suite 1800, Philadelphia, PA 19103-2899. Customer Service Office: 6277 Sea Harbor Drive, Orlando, FL 32887-4800. Subscription prices are $146.00 per year for US individuals, $253.00 per year for US institutions, $76.00 per year for US students and residents, $173.00 per year for Canadian individuals, $292.00 per year for Canadian institutions, $184.00 per year for international individuals, $292.00 per year for international institutions and $92.00 per year for Canadian and foreign students/residents. To receive student/resident rate, orders must be accompanied by name of affiliated institution, date of term, and the *signature* of program/residency coordinator on institution letterhead. Orders will be billed at individual rate until proof of status is received. Foreign air speed delivery is included in all *Clinics* subscription prices. All prices are subject to change without notice.

**POSTMASTER:** Send address changes to *Veterinary Clinics of North America: Exotic Animal Practice*; Elsevier Periodicals Customer Service, 6277 Sea Harbor Drive, Orlando, FL 32887-4800. **Customer Service: 1-800-654-2452 (US). From outside of the US, call 1-407-345-1000.**

*Veterinary Clinics of North America: Exotic Animal Practice* is covered in *Index Medicus*.

Printed in the United States of America.

# CYTOLOGY

## CONSULTING EDITOR

**AGNES E. RUPLEY, DVM,** Diplomate, American Board of Veterinary Practitioners–Avian; Director and Chief Veterinarian, All Pets Medical & Laser Surgical Center, College Station, Texas

## GUEST EDITOR

**MICHAEL M. GARNER, DVM,** Diplomate, American College of Veterinary Pathologists; Northwest ZooPath, Monroe, Washington

## CONTRIBUTORS

**A. RICK ALLEMAN, DVM, PhD,** Diplomate, American Board of Veterinary Practitioners; Diplomate, American College of Veterinary Pathologists; Associate Professor of Clinical Pathology, University of Florida, College of Veterinary Medicine, Gainesville, Florida

**GREGORY D. BOSSART, VMD, PhD,** Senior Scientist and Head of Pathology, Division of Marine Mammal Research and Conservation, Center for Coastal Research–Marine Mammal Research and Conservation Program, Harbor Branch Oceanographic Institution, Fort Pierce, Florida

**TERRY W. CAMPBELL, MS, DVM, PhD,** Associate Professor of Zoological Medicine, Department of Clinical Sciences, College of Veterinary Medicine and Biomedical Sciences; Program Chair and Chief of Services in Zoological Medicine, Veterinary Medical Center, Colorado State University, Fort Collins, Colorado

**MICHAEL M. GARNER, DVM,** Diplomate, American College of Veterinary Pathologists; Northwest ZooPath, Monroe, Washington

**JULI D. GOLDSTEIN, DVM,** Consulting Veterinarian, Division of Marine Mammal Research and Conservation, Harbor Branch Oceanographic Institution, Fort Pierce, Florida

**CARLES JUAN-SALLÈS, DVM,** Diplomate, American College of Veterinary Pathologists; and ConZOOlting Wildlife Management, Samalús (Barcelona), Spain

**EMILY K. KUPPRION, VMD,** Animal Emergency Hospital–Volusia, Ormond Beach, Florida

**KENNETH S. LATIMER, DVM, PhD,** Diplomate, American College of Veterinary Pathologists; Professor and Director of Clinical Pathology Laboratory, Department of Pathology, College of Veterinary Medicine, University of Georgia, Athens, Georgia

**GREGORY A. LEWBART, MS, VMD,** Diplomate, American College of Zoological Medicine; Professor of Aquatic Animal Medicine, Department of Clinical Sciences, College of Veterinary Medicine, North Carolina State University, Raleigh, North Carolina

**ALLAN P. PESSIER, DVM,** Diplomate, American College of Veterinary Pathologists; Associate Pathologist, Wildlife Disease Laboratories, Conservation and Research for Endangered Species, Zoological Society of San Diego, San Diego, California

**PAULINE M. RAKICH, DVM, PhD,** Diplomate, American College of Veterinary Pathologists; Associate Professor and Diagnostic Pathologist, Athens Veterinary Diagnostic Laboratory, College of Veterinary Medicine, University of Georgia, Athens, Georgia

**DRURY REAVILL, DVM,** Diplomate, American Board of Veterinary Practitioners-Avian; Diplomate, American College of Veterinary Pathologists; Zoo/Exotic Pathology Service, Citrus Heights, California

**HELEN ROBERTS, DVM,** 5 Corners Animal Hospital, Orchard Park, New York

**KIMBERLY SCHMIDT, BS,** Veterinary Student, College of Veterinary Medicine, University of Illinois at Urbana–Champaign, Urbana, Illinois

**RENÉ A. VARELA, MS, VMD,** Director of Veterinary Services, Ocean Embassy, Orlando, Florida

**ARNAUD VAN WETTERE, DVM, MS,** Anatomic Pathology Resident, Department of Population Health and Pathobiology, College of Veterinary Medicine, North Carolina State University, Raleigh, North Carolina

# CYTOLOGY

# CONTENTS

**Preface** xi
Michael M. Garner

**Basics of Cytology and Fluid Cytology** 1
Terry W. Campbell

    Cytology as a diagnostic tool has played a major role in the management of diseases affecting domestic mammals for over 20 years. It has also become a valuable diagnostic tool in the evaluation of nondomestic or the so-called "exotic" animal patients, such as small mammals and the lower vertebrates. Common cytologic specimens used to evaluate the exotic animal patient include aspirates of masses and organs, imprints of biopsy material, tracheal wash samples, aspiration of abdominal or coelomic fluid, and fecal smears. In general, the same cytologic sample collection and preparation techniques used for domestic mammals also apply to exotic animal patients. The interpretation of the cytology specimen is generally the same as that of domestic mammals.

**Cytologic Diagnosis of Diseases of Rabbits, Guinea Pigs, and Rodents** 25
Michael M. Garner

    This article reviews the diseases most amenable to cytologic diagnosis in clinical small mammal practice. Diseases of pet rabbits, guinea pigs, rodents, chinchillas, sugar gliders, and hedgehogs are addressed. The small size of these patients, risk of anesthesia and invasive surgery, and cost factors make small mammals ideal patients for cytologic evaluation when applicable; however, surprisingly few reports exist in the literature, and no other reviews of cytology in these species exist. Much of the data in this article is derived from case submissions to Northwest ZooPath, and disease presentations that seem to be common in this group of animals are emphasized. Diseases of the skin are particularly well represented, especially tumors and inflammatory processes.

## Cytologic Diagnosis of Diseases of Hedgehogs 51
Carles Juan-Sallés and Michael M. Garner

This article focuses on neoplastic diseases because they may be the most frequent disease processes in captive hedgehogs according to the literature and authors' case files and the most common cases submitted for cytologic diagnosis in these species, particularly the African hedgehog (*Atelerix albiventris*).

## Cytologic Diagnosis of Diseases of Ferrets 61
Pauline M. Rakich and Kenneth S. Latimer

Cytology is a useful, rapid, inexpensive diagnostic technique that is particularly suitable for ferrets because of their small size and readily accessible organs and tissues. This article begins with a brief discussion of general cytologic information. The remainder of the article concentrates on the cytologic features of common diseases that affect ferrets.

## Evaluation of Cetacean and Sirenian Cytologic Samples 79
René A. Varela, Kimberly Schmidt, Juli D. Goldstein, and Gregory D. Bossart

Cytology is a fundamental part of marine mammal veterinary medicine that is involved in preventive medicine programs in captive animals and in the health assessment of wild populations. Marine mammals often exhibit few clinical signs of disease; thus, the cost-effective and widely accessible nature of cytologic sampling renders it one of the most important diagnostic procedures with these species. Many of these mammals are endangered, protected, and located in developing nations in which resources may be scarce. This article can be used as a field guide to advise a veterinarian, biologist, or technician working with cetaceans or sirenians. A simplistic cost-effective staining technique is used, which is ideal for situations in which funds, facilities, or time may be a limiting factor in clinical practice.

## Avian Cytology 131
Kenneth S. Latimer and Pauline M. Rakich

An overview of avian cytology is presented, discussing more common abnormalities that are encountered in routine clinical practice. The general cytologic features of inflammatory, infectious, and neoplastic lesions are described. The remainder of the article covers major cytologic abnormalities by anatomic site of origin of the specimens.

### Cytologic Diagnosis of Diseases in Reptiles 155
A. Rick Alleman and Emily K. Kupprion

The cytologic evaluation of samples obtained from reptile patients may provide invaluable diagnostic information to the clinician. The following article is directed toward providing information regarding the techniques used to obtain samples, discussion of sample types, and guidelines for the cytologic classification of the materials collected from tissue lesions and body fluids.

### Cytologic Diagnosis of Disease in Amphibians 187
Allan P. Pessier

Cytology is an inexpensive yet powerful diagnostic tool that allows for rapid diagnosis of many common disease conditions in amphibian patients. Although the emphasis of this article is on infectious diseases, there is great potential for application of cytologic diagnosis to variety of medical conditions as the knowledge base in amphibian medicine and pathology continues to grow. Routine methods used that may fall under the umbrella of cytology range from wet mount examination of skin scrapings (or gill biopsies of larvae) to examination of stained impression smears. Routine Romanowsky's-type stains work well for amphibian samples. Preparation of multiple smears is always recommended to allow for use of special staining procedures.

### Diagnostic Cytology of Fish 207
Drury Reavill and Helen Roberts

Cytology is an essential part of a diagnostic workup in cases of aquatic animal diseases. It is simple to perform, inexpensive, and can yield quick and valuable results. External parasites, bacterial and fungal diseases, and gastrointestinal infestations are easily determined with wet mount cytology. Because of the relatively small number of nonlethal diagnostic techniques available for aquatic species, cytologic testing should be considered in every case. Early diagnosis can lead to more effective treatment plans, ensuring a better prognostic outcome in our patients.

### Cytologic Diagnosis of Diseases of Invertebrates 235
Arnaud Van Wettere and Gregory A. Lewbart

Invertebrate medicine in the context of an exotic or zoo animal veterinary practice is in its infancy. Establishment of species-specific reference values and evaluation of the effectiveness of cytology for diagnosis of specific diseases are necessary. Despite the lack of normal reference parameters for most species encountered in clinical practice, important information may be obtained from cytologic examination of tissue imprints, aspirates, scrapings, and hemolymph. This information may be essential to establish a specific diagnosis, focus investigations, and influence treatments. It is hoped that this article stimulates veterinarians who work with invertebrates to use diagnostic cytology and disseminate the results of their experience.

**Index**     **255**

## FORTHCOMING ISSUES

May 2007
### Critical Care
Marla Lichtenberger, DVM, DAVECC
*Guest Editor*

September 2007
### Neurology
Lisa A. Tell, DVM and
Marguerite Knipe, DVM, *Guest Editors*

## RECENT ISSUES

September 2006
### Case Reports: The Front Line in Exotic Medicine
Robert J.T. Doneley, BVSc, FACVSc,
*Guest Editor*

May 2006
### Common Procedures
Chris Griffin, DVM, Dipl. ABVP–Avian
*Guest Editor*

January 2006
### Renal Disease
M. Scott Echols, DVM, Dipl. ABVP–Avian
*Guest Editor*

---

**The Clinics are now available online!**

Access your subscription at
**www.theclinics.com**

VETERINARY
CLINICS
Exotic Animal Practice

# Preface

Michael M. Garner, DVM, DACVP
*Guest Editor*

The application of cytologic examination to the diagnostic work-up was more or less in its infancy when I entered veterinary school in the early 1980s. I remember a brief section in the clinical pathology course in the junior year, followed by a few lectures during a week on rounds in the senior year. Mostly, we looked at abnormal blood cells, or vaginal smears, or "gorp" floating around in urine sediments, or maybe a lymphoid malignancy or mast cell tumor. The discipline seemed limited, and perhaps prone to imaginative interpretation.

On the near horizon, however, useful tools and energetic minds were emerging; such textbooks as Tyler and Cowell's *Diagnostic Cytology and Hematology of the Dog and Cat*, and quick Romanowsky's staining techniques, could be used in-house. From there the discipline in domestic species seemed to blossom. A young clinician so inclined might stick a needle into any tissue, lesion or not, and come up with a diagnosis. (I remember describing to myself in great detail the cell populations in the vomitus of a dog with parvo.) For some of us, cytology became a cheap, readily available panacea to the doldrums of routine practice. Of course, it was good to get histologic confirmation of the lesion at the local pathology laboratory, always for me a humbling experience. I am sure the wise old pathologists suffering through my cytologic descriptions enjoyed a hearty laugh or two at my expense.

From those early days, cytopathology has emerged as a staple in the diagnostic work-up, and not just for domestic species, but also for exotic pets and zoo or wildlife species. This issue of *Veterinary Clinics of North America: Exotic Animal Practice* taps the experiences and talents of several of the leading experts in the field of exotic species pathology, and although not entirely comprehensive, provides a broad review of the cytology of exotic species. It has been my great pleasure to work with these folks in the development of this volume, and I hope readers find this issue as useful in their practice as I do in mine.

Michael M. Garner, DVM, DACVP
*Northwest ZooPath*
*654 West Main*
*Monroe, WA 98296, USA*

*E-mail address:* zoopath@aol.com

# Basics of Cytology and Fluid Cytology
Terry W. Campbell, MS, DVM, PhD[a,b],*

[a]Department of Clinical Sciences, College of Veterinary Medicine and Biomedical Sciences, Colorado State University, 300 West Drake Road, Fort Collins, CO 80523, USA
[b]Veterinary Medical Center, Colorado State University, 300 West Drake Road, Fort Collins, CO 80523, USA

Cytology as a diagnostic tool has played a major role in the management of diseases affecting domestic mammals for over 20 years. It has also become a valuable diagnostic tool in the evaluation of nondomestic or the so-called "exotic" animal patients, such as small mammals (rabbits, rodents, ferrets, hedgehogs, and so forth) and the lower vertebrates, especially birds and reptiles. Common cytologic specimens used to evaluate the exotic animal patient include aspirates of masses and organs, imprints of biopsy material, tracheal wash samples, aspiration of abdominal or coelomic fluid, and fecal smears. Wash or aspirate samples of the crop (ingluvies) of birds, the stomach of reptiles, and lungs of reptiles are also commonly evaluated cytology specimens. Standard veterinary cytodiagnosis methods have had limited use in the evaluation of aquatic patients, such as amphibians and fish, where diagnostic microscopy using wet mount preparations have prevailed in the management of those animals. In general, the same cytologic sample collection and preparation techniques used for domestic mammals also apply to exotic animal patients. Likewise, the interpretation of the cytology specimen is generally the same as that of domestic mammals.

Romanowsky-type stains (Wright's, Giemsa, Wright's Giemsa, Leishman's, and May-Grünwald-Giemsa stains) are the most commonly used stains in veterinary cytology and are especially useful in the clinical practice setting. Wright's stain (Wright's Stain Solution, Fisher Scientific, Pittsburgh, PA) is the standard stain used to evaluate blood films for hematology studies. Other Romanowsky-type stains used either alone or in combination include the so-called "quick" or "stat" stains, which have a simple staining procedure, rapid staining time, and generally good staining quality.

---

* Department of Clinical Sciences, College of Veterinary Medicine and Biomedical Sciences, Colorado State University, 300 West Drake Road, Fort Collins, CO 80523.
 *E-mail address:* twc@colostate.edu

A general impression of the quality of the cytologic sample is made using scanning (×40 or ×100) and low (×200) magnifications. At these magnifications, the examiner is able to estimate the smear cellularity; examine cellular aggregates; identify large infectious agents (eg, yeast and fungal elements); and determine the best locations for examination at higher magnifications (monolayers of cells). Higher magnification (ie, high-dry, ×400 or ×500, and oil immersion, ×500 and ×1000) are used to examine cell structure, bacteria, and other small structures, such as cellular inclusions.

The emphasis of cytodiagnosis is placed on the general appearance of the cells and their nuclei. Important cellular features include sample cellularity, cellular distribution, size and shape of the various cells, and the cytoplasmic appearance of the cells. Important nuclear features to note include the size, shape, and position of the nucleus within the cell; the number of nuclei present; the chromatin pattern present within the nucleus; the appearance of the nucleoli; and the presence of mitotic figures.

The size of the cell nucleus is often evaluated in relation to the amount of cytoplasm. The nuclear to cytoplasmic ratio (N:C ratio) is an indication of the relationship between the size of the nucleus and the cytoplasmic volume. Mature, well-differentiated epithelial cells tend to have small nuclei and a low N:C ratio, whereas malignant, poorly differentiated epithelial cells frequently have large nuclei and a high N:C ratio. The shape of the nucleus is often related to the shape of the cell; however, this does not hold true for all cells. The nucleus is frequently centrally located in epithelial cells but may be eccentric if displaced by secretory granules or vacuoles. Most normal cells have a single nucleus; however, some normal cells may contain multiple nuclei. For example, some hepatocytes are binucleated and osteoclasts are multinucleated. The nuclear chromatin patterns commonly noted include uniformly finely granular chromatin, finely granular chromatin with irregular distribution, uniformly coarsely granular chromatin, and coarsely granular chromatin with irregular distribution. The finely granular chromatin patterns generally indicate nuclear immaturity. Some nuclei (eg, nuclei of mature lymphocytes) often contain one or more large, prominent, chromatin clumps called chromocenters or false nucleoli. Mitotic figures indicate cells undergoing active division, which can be a normal cellular feature of some tissues, such as the bone marrow and liver, when they are found in low numbers. High numbers of mitotic figures or abnormal-appearing mitotic figures indicate abnormalities, such as neoplasia. Nucleoli appear as clear, circular spaces inside the nucleus of cells stained with Wright's stain. Many normal cells often contain one or more small nucleoli. Large or irregular nucleoli are considered abnormal. Nuclei with numerous nucleoli (ie, greater than five) are also considered to be abnormal.

The characteristics of the noncellular background of a cytologic specimen should not be overlooked because it may provide clues to the nature of the material being examined. For example, smears containing many secretory cells may have heavy background material because of the accumulation of a secretory product. A finely granular background in smears of inflammatory exudates stained with new methylene blue suggests an increase in protein content. A coarsely granular background is present in smears containing high amounts of mucopolysaccharides (eg, mucin) when stained with Wright's stain. The presence of bacteria, crystals, lipid droplets, nuclear materials from ruptured cells, and foreign material (eg, plant cells, pollen, talcum, or starch crystals) should also be noted. Also, cytologic samples often contain variable amounts of peripheral blood. Excessive peripheral blood contamination of the sample dilutes and masks the diagnostic cells, making the cytologic interpretation difficult.

Cells observed in cytologic specimens can generally be classified by origin into one of four major tissue groups: (1) epithelial-glandular, (2) connective, (3) nervous tissue, or (4) hemic tissue. Epithelial cells are usually polygonal, spherical, cuboidal, or columnar; have distinct cell margins; exhibit an abundant cytoplasm and a small nucleus; and exfoliate in cell aggregates (Fig. 1). Cells from connective tissue tend to be long and thin (spindle-shaped and elongated); have indistinct cell margins; have variable cytoplasmic volumes and nuclear shapes depending on their origin; and exfoliate as individual cells (Fig. 2). Cells derived from the nervous system are rarely seen on cytologic specimens and appear as deeply basophilic stellate cells with cytoplasmic projections.

Cells derived from hemic (blood and blood-forming) tissue are found in peripheral blood; bone marrow; and ectopic hematopoietic sites in various organs (eg, spleen and liver). It is important that the cytologist become familiar with cells from hemic tissue because many of these cells are important features of various cytologic responses. Also, hemic cells are a frequent contaminate of the cytologic specimen and should be recognized as such. The

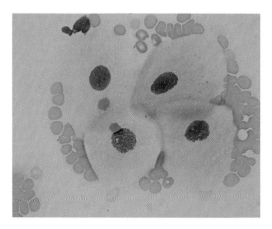

Fig. 1. Intermediate squamous epithelial cells from the oral cavity of a tiger (Wright's stain, original magnification ×1000).

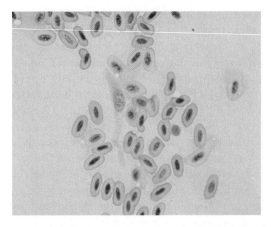

Fig. 2. A spindle-shaped cell indicative of a mesenchymal cell (Wright's stain, original magnification ×1000).

hemic cells include erythrocytes, leukocytes, and either platelets in the case of mammals or thrombocytes in the case of lower vertebrates.

Mammalian erythrocytes are small, anucleate, round, biconcave in most species, and often appear to have a central area of pallor caused by the biconcavity of the cell (Fig. 3). The cytoplasm appears orange-pink on Romanowsky-stained (ie, Wright stained) blood films. The much larger mature nucleated erythrocytes of lower vertebrates, such as birds, reptiles, amphibians, and fish, generally appear as flattened elliptical cells with an elliptical, centrally positioned nucleus (Fig. 4).

Leukocytes in the blood include granular (granulocytes) and agranular leukocytes. The agranular leukocytes are also referred to as "mononuclear leukocytes" and include lymphocytes and monocytes. Lymphocytes and

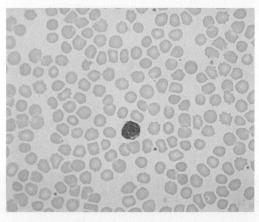

Fig. 3. A normal small mature lymphocyte, erythrocytes, and platelets in the blood film of a rabbit (Wright's stain, original magnification ×1000).

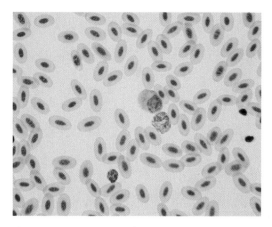

Fig. 4. Normal erythrocytes, a monocyte, a heterophil, and thrombocytes in the blood film from a bird (Wright's stain, original magnification ×1000).

monocytes have an identical morphology among all animals. A lymphocyte or monocyte found in the blood film of a mammal also resembles those found in blood films of birds, reptiles, amphibians, and fish. The granulocytes, however, may appear differently among the species. They are further classified as neutrophils or heterophils, eosinophils, and basophils. The leukocytes can participate in inflammatory responses and are found in inflammatory lesions.

Lymphocytes are typically round cells with a round, occasionally slightly indented, centrally or slightly eccentrically positioned nucleus and the nuclear chromatin is heavily clumped or reticulated in mature cells (see Fig. 3). The cytoplasm is typically scant (giving lymphocytes a high N:C); homogenous; weakly basophilic (pale blue); and lacks vacuoles or granules.

Monocytes are typically the largest leukocyte present in the blood film of animals (see Fig. 4). Monocytes vary in shape from round to amoeboid. The nucleus also varies in shape from round to lobed, and is relatively pale with less chromatin clumping as compared with a lymphocyte nucleus. The cytoplasm is abundant, blue-gray in color, and may appear slightly opaque. Vacuoles or fine dust-like eosinophilic granules may be present. Monocytes exhibit phagocytic activity and migrate into tissues to become macrophages. Monocytes and macrophages possess biologically active chemicals involved in inflammation mediation and the destruction of invading organisms.

Neutrophils contain numerous small granules that vary from colorless, pale, or dark staining among different species of mammal. Neutrophils of mice, rats, and chinchillas typically have a colorless cytoplasm, although some dust-like red granules may be present within the cytoplasm, which may cause the cell to stain diffusely pink with Romanowsky stains (Fig. 5). The neutrophils of rabbits, guinea pigs, hamsters, and gerbils are often referred to as "heterophils" or "pseudoeosinophils" because they

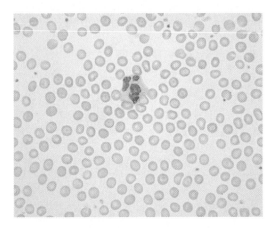

Fig. 5. A normal neutrophil (heterophil), erythrocytes, and platelets in the blood film of a chinchilla (Wright's stain, original magnification ×1000).

contain granules that stain eosinophilic in color with Romanowsky stains (Fig. 6).

The heterophil of lower vertebrates is functionally equivalent to the mammalian neutrophil. In general, heterophils tend to be round, although their shape may be distorted during blood film preparation. The nucleus of the mature heterophil either lacks distinct lobes (ie, most reptiles) or is lobed in some species (ie, birds and some lizards) with coarse, clumped, purple-staining chromatin (Figs. 7 and 8). The nucleus is often partially hidden by the cytoplasmic granules. The cytoplasm of normal mature heterophils appears colorless and contains granules that stain an eosinophilic color (dark orange to brown-red) with Romanowsky stains. The granules can

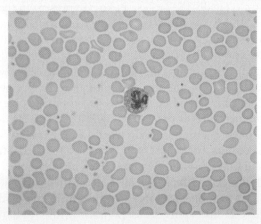

Fig. 6. A rabbit heterophil, erythrocytes, and platelets (Wright's stain, original magnification ×1000).

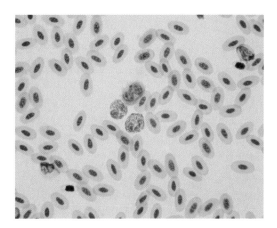

Fig. 7. Two heterophils, a monocyte, and two lymphocytes among normal erythrocytes in the blood film of a bird (Wright's stain, original magnification ×1000).

be affected by the staining process, and may appear atypical (ie, poorly stained, partially dissolved, or fused) in some cases. The shape, size, and general morphology of the cytoplasmic granules may vary among the different species of lower vertebrates. Typically, the cytoplasmic granules appear elongated (rod or spiculated in shape), but may also appear oval to round.

Mammalian eosinophils contain large eosinophilic cytoplasmic granules and are particularly numerous in the peripheral blood of mammals when antigens are continually being released, as occurs in parasitic disease (especially those involving larvae of helminthes) and allergic reactions (especially those associated with mast cell and basophil degranulation) (Fig. 9). Less is known about the function of cells in the peripheral blood of lower

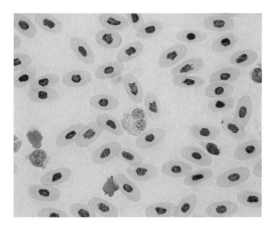

Fig. 8. A heterophil, lymphocyte, thrombocyte, and erythrocytes in the blood film of a lizard (*Iguana iguana*) (Wright's stain, original magnification ×1000).

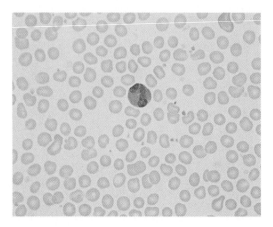

Fig. 9. An eosinophil, erythrocytes, and platelets in the blood film of a guinea pig (Wright's stain, original magnification ×1000).

vertebrates that are called "eosinophils." Eosinophils from these animals can resemble heterophils. In comparison, eosinophils usually have a darker-staining nucleus, a blue cytoplasm, and the tinctorial quality of the granules varies from heterophils in the same stained blood film. The appearance of the granules of eosinophils of lower vertebrates may also vary from the typical eosinophilic color to granules that appear colorless; pale blue; or as pale, large, swollen, round structures (Fig. 10) [1].

Basophils of mammals and lower vertebrates are easily identified by their characteristic cytoplasmic granules that are strongly basophilic on Romanowsky-stained blood films. Basophils of mammals tend to have

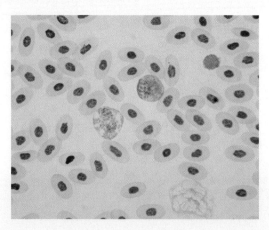

Fig. 10. A heterophil, an eosinophil (cell with blue cytoplasmic granules), and erythrocytes in the blood film of a lizard (*Iguana iguana*) (Wright's stain, original magnification ×1000).

lobed nuclei; however, those of lower vertebrates tend to be nonlobed (Fig. 11) [2,3].

Mammalian platelets are flat anucleated disks of cytoplasm that contain cytoplasmic organelles. They are derived as cytoplasmic fragments with variable amounts of small purple granules on Romanowsky-stained blood films and arise from megakaryocytes within the bone marrow. Platelets tend to be round, but can vary slightly in shape and size (see Figs. 3, 5, 6, and 9). Platelets are involved in the clotting process and are responsible for the initial hemostatic plug to prevent hemorrhage after vascular injury to the microcirculation. Because of this function, they are often found in clumps on blood films.

The lower vertebrates (birds, reptiles, amphibians, and fish) have true nucleated cells called "thrombocytes" that function in a similar manner to mammalian platelets. Thrombocytes are typically small round to oval cells (smaller than erythrocytes) with a round to oval nucleus, high N:C ratio, and colorless to pale gray reticulated cytoplasm that often contains red granules at one pole of the cell (see Figs. 4, 8, and 11). Activated thrombocytes (those participating in hemostasis) appear as aggregated clusters of cells with decreased cytoplasmic volume; irregular cytoplasmic margins (often indistinct); and vacuoles. Activated forms may also have fusiform shapes and small eosinophilic granules [4].

The thrombocytes of birds and perhaps other lower vertebrates are capable of phagocytosis and may participate in removing foreign materials from the blood. Although they are less phagocytic than the heterophils, they may act as a nonspecific "scavenging" phagocyte, capable of clearing a wide range of foreign objects, including bacteria [5]. Although thrombocytes have the capacity of phagocytic activity, they are rarely found as part of the inflammatory response as viewed in a cytologic specimen.

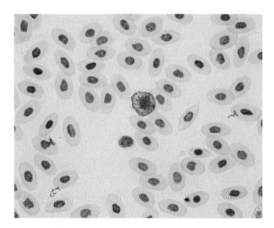

Fig. 11. A basophil, a thrombocyte, and normal erythrocytes in the blood film of a lizard (*Iguana iguana*). (Wright's stain, original magnification ×1000).

## Inflammation

Inflammation occurs whenever chemotactic factors for inflammatory cells, such as vasoactive amines (ie, 5-hydroxy-tryptamine from mast cell granules), are released. The most common causes are microbes and their toxins, physical and chemical trauma, death of cells from circulatory insufficiency, and immune reactions. The rate of the inflammatory response is temperature dependent in poikilotherms, such as reptiles, amphibians, and fish, whereas the rate of response in endotherms, such as mammals and birds, is more consistent and predictable.

The inflammatory process begins with the acute phase. Leukocytes actively migrate into the affected tissue. The degree of the cellular migration depends on the stimulus, which is particularly marked in certain bacterial infections. The cells that leave the blood include neutrophils or heterophils depending on the species, monocytes, lymphocytes, and thrombocytes of lower vertebrates. Neutrophils or heterophils are typically the first cells to arrive at the scene where they destroy ingested organisms. Monocytes in circulation in conjunction with local tissue macrophages (which have multiplied at the site of inflammation) begin to phagocytize tissue debris and infectious agents as the inflammation becomes established. Monocytes and macrophages are stimulated by the cell-mediated immune response and are the dominant inflammatory cell in most cellular inflammatory responses and are capable of developing into epithelioid and multinucleated giant cells. As the inflammatory process continues and becomes chronic, granulomas may form as the macrophages develop into layers that resemble epithelium and is the reason for the term "epithelioid cells." As the lesion matures, fibroblasts proliferate and begin to lay down collagen. These proliferating fibroblasts appear large compared with the small densely staining fibroblasts of normal fibrous tissue. Lymphocytes appear within the stroma and participate in the cell-mediated immune response. Fusion of macrophages into giant cells occurs in association with material that is not readily digested by macrophages. The results of acute inflammation may be complete resolution, development of an exudative or necrotic lesion with continuation of the inflammatory response, or progression to chronic inflammation.

The inflammatory response of animals can be classified as either neutrophilic or heterophilic, eosinophilic, mixed cell, or macrophagic depending on the predominant cell type. Neutrophilic inflammation, also referred to as "purulent" or "suppurative" inflammation, of mammals and heterophilic inflammation of birds and reptiles is represented by a predominance of neutrophils or heterophils (greater than 80% of the inflammatory cells)] in the cytologic sample (Figs. 12 and 13). Heterophil granules in cytologic specimens tend to lose their normal rod-shaped appearance and either appear more rounded or absent.

Mammalian neutrophils and heterophils of lower vertebrates are highly phagocytic, have considerable bactericidal activity, and actively participate

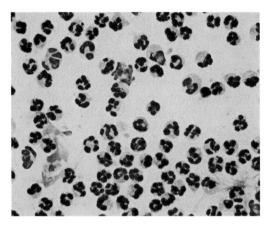

Fig. 12. Neutrophilic inflammation in an aspirate from a guinea pig (Wright's stain, original magnification ×1000).

in inflammatory lesions [1,6–9]. Less is known about the function of the heterophils and neutrophils of amphibians and fish.

Neutrophilic and heterophilic inflammation are classified by the presence or absence of degeneration of these granulocytes [10]. The nuclear features of degenerate heterophils and neutrophils include swelling, karyorrhexis, and karyolysis. Karyorrhexis or rupture of the nuclear membrane and fragmentation of the nuclear chromatin indicates the end stage of cell death and is represented by multiple pyknotic nuclear segments (representing nuclear fragmentation) in the center of the cell. Karyolysis is represented by a nucleus that has lost its basophilia and appears swollen with poorly defined, homogenous pink chromatin with Romanowsky stains. The cytoplasmic

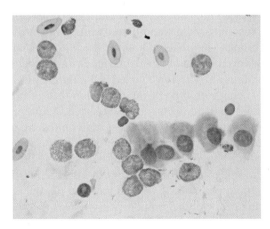

Fig. 13. A heterophilic inflammation in the smear made from a conjunctival scraping from a bird (Wright's stain, original magnification ×1000).

features of degenerate heterophils and neutrophils include increased basophilia, vacuolization, and with heterophils, varying degrees of degranulation. Degenerate heterophils and neutrophils represent rapid cell death and suggest the presence of toxins, such as bacterial toxins, in the microenvironment [11]. Some aerobic bacteria and fungi cause degenerative cell changes, whereas many anaerobic bacteria and mycoplasma initiate chemotaxis, but do not affect the morphology of the neutrophils or heterophils [12].

Nuclear pyknosis indicates a slow progressive degeneration of a cell in a nontoxic environment and may represent the natural aging of the cell. Pyknosis is characterized by nuclear shrinkage causing the chromatin to become dense and deeply basophilic. Pyknotic nuclei appear as a single round mass with an intact nuclear membrane.

Heterophilic inflammation usually indicates an acute phase of the inflammatory response in birds and reptiles [13]. Basophils and thrombocytes may also participate in the early inflammatory response [14,15]. In about 7 days, with the macrophage involvement, the characteristic heterophilic granuloma develops [6]. Giant cell formation is a common occurrence in avian and reptilian inflammatory lesions because the necrotic tissue stimulates a foreign body–like reaction. Unlike mammalian giant cell formation, the presence of giant cells in avian inflammatory lesions does not necessarily suggest chronicity.

As with other vertebrates, acute inflammation in fish is initiated by the action of vasoactive amines and cell breakdown products released by the tissue damage on the microcirculation of the remaining tissue [16]. Neutrophils are less significant in fish inflammation than other vertebrates, where they are seen in early stages of inflammation, but not in the later stages. Typical abscesses found in other vertebrates are uncommon in fish.

Septic inflammation is indicated by the presence of intracellular bacteria. Bacteria that have been phagocytized by leukocytes often appear within vacuoles called phagosomes, membrane-bound vesicles formed by invagination of the cell membrane. Most bacteria stain blue with Romanowsky stains.

Because of the rapid influx of macrophages (within a few hours) and lymphocytes into inflammatory lesions, mixed cell inflammation (pyogranulomatous) is the most common type of inflammation seen in birds and reptiles. Mixed cell inflammation indicates an established, active inflammatory lesion. Mixed cell inflammation is typically represented by a predominance of neutrophils or heterophils (greater than 50% of the inflammatory cells) with an increased number of mononuclear leukocytes (Fig. 14). Neutrophils and heterophils in mixed-cell inflammatory lesions are usually normal, that is nondegenerate in appearance. Lymphocytes and plasma cells are often associated with acute heterophilic granulomas, whereas the presence of epithelioid cells (macrophages that contain no vacuoles or phagocytized material) and connective tissue cells (ie, fibroblasts) suggest chronic granulomas. Frequently, the epithelial and mesenchymal cells adjacent to inflammatory

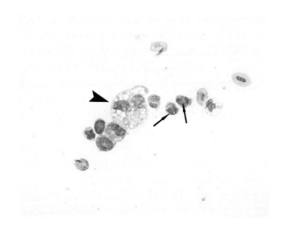

Fig. 14. A fluid sample obtained from the coelomic cavity from a female bird with egg-related coelomitis reveals heterophils (*arrows*), macrophages, and a multinucleated giant cell (*arrowhead*) indicative of a mixed cell inflammation (Wright's stain, original magnification ×1000).

lesions proliferate resulting in the presence of these types of cells showing features of tissue hyperplasia.

Macrophagic (histiocytic) inflammation in mammals is suggestive of chronic inflammation and is often seen with foreign body reactions and mycobacterial infections [10]. Macrophagic inflammation may have a different pathogenesis than heterophilic and mixed cell inflammation in lower vertebrates, such as birds and reptiles. Macrophagic inflammation is indicated by a predominance of macrophages (greater than 50% of the inflammatory cells) in the cytologic sample (Fig. 15). Large activated macrophages resembling epithelial cells (epithelioid macrophages) that later develop into multinucleated giant cells, apparently responding to necrotic tissue, are a feature

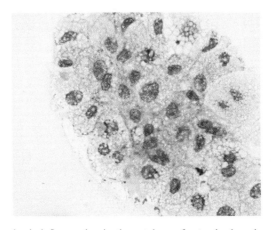

Fig. 15. A macrophagic inflammation in the cytology of a tracheal wash sample from eagle (Wright's stain, original magnification ×1000).

of this type of inflammation. Macrophagic inflammation is common in certain diseases, such as mycobacterial and chlamydial infections and cutaneous xanthomatosis. Multinucleated giant cells are found in granulomatous lesions in inflammatory diseases, such as avian tuberculosis [17]. Areas of macrophagic inflammation and heterophilic inflammation can occur together as macrophages respond to necrotic materials. Depending on where the sample is obtained, the cytodiagnosis may be a macrophagic inflammation.

Melanomacrophages are pigmented phagocytic cells found in most fish, amphibians, and reptiles (Fig. 16) [18,19]. Melanomacrophages are aggressive phagocytic cells that readily consume fungi, bacteria, parasites, mycobacteria, foreign bodies, and cell debris, including hemoglobin breakdown products [20]. Because melanomacrophages function more affectively at lower temperatures compared with mammalian macrophages, they are likely to play a key roll in controlling infections during times of extended hypothermia in ectothermic animals [20]. Melanomacrophages may elicit a granulomatous, encapsulating response to infectious agents and foreign material [19]. Chronically diseased fish, amphibians, and reptiles exhibit an increase in the size and number of melanomacrophage centers in their tissues [21].

Lymphocytic and plasmacytic infiltration is indicated by increased numbers of lymphocytes and plasma cells in the cellular response [10]. Conditions that are often associated with this type of cellular response include early viral infections, immune-mediated disorders, and chronic inflammation. The lymphocyte population is composed of small and intermediate-sized mature lymphocytes and plasma cells. Plasma cells are large, round-to-oval lymphocytes with an abundant, deeply basophilic cytoplasm (Figs. 17 and 18). The nucleus is eccentrically located and appears mature. A prominent Golgi apparatus is found adjacent to the nucleus.

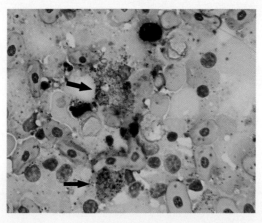

Fig. 16. Heterophils, erythrocytes, and melanomacrophages (*arrows*) in a liver aspirate from a reptile (Wright's stain, original magnification ×1000).

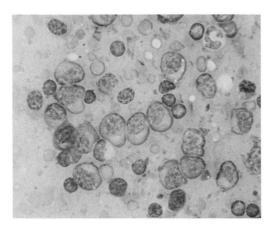

Fig. 17. The aspirate of a splenic mass from a ferret with a multiple myeloma reveals numerous plasma cells (Wright's stain, original magnification ×1000).

In mammals and perhaps lower vertebrates, an increased number of eosinophils (10% of the inflammatory cells or greater) in the inflammatory response is indicative of an eosinophilic inflammation. Eosinophils are particularly numerous when antigens are continually being released, as in parasitic disease. Eosinophilic inflammation is often associated with hypersensitivity or allergic reactions, parasites, mast cell tumors, and eosinophilic granulomas. Eosinophilic inflammation is rare in birds and reptiles.

### Tissue hyperplasia and benign neoplasia

Based on cytomorphology, tissue hyperplasia and benign neoplasia are indistinguishable. Cells representative of tissue hyperplasia or benign

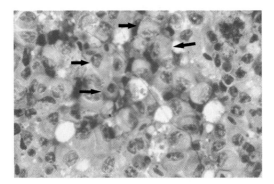

Fig. 18. A splenic aspirate from a parrot with chlamydiosis reveals numerous plasma cells (*some represented by arrows*) and represents plasma cell hyperplasia (Wright's stain, original magnification ×1000).

neoplasia exhibit uniformity in overall appearance, have an increased cytoplasmic basophilia with pale vesicular nuclei, and are similar in size with similar N:C ratios and nuclear features [22]. Cells associated with cellular hyperplasia may exhibit an increase in normal mitotic figures, which is indicative of the proliferative nature of the tissue. Cells suggestive of hyperplasia of epithelial and connective tissue often occur in cytologic specimens of long-standing chronic inflammation (Fig. 19).

## Malignant neoplasia

The cytologic criteria used for the diagnosis of malignant neoplasia can be divided into four categories: (1) general cellular, (2) cytoplasmic, (3) nuclear, and (4) structural features. General cellular features of malignant neoplasia refer to the appearance of the cell population present in the cytology sample. This includes the presence of a monomorphic cell population in the absence of inflammation, pleomorphism among noninflammatory cells with an apparent common origin, increased cellularity in samples from tissues that normally provide low cellular samples, and the appearance of cells that are foreign to the tissue being sampled.

The appearance of the cytoplasm may also aid in the diagnosis of malignant neoplasia. Cytoplasmic features suggestive of malignant neoplasia include basophilia (indicating increased RNA activity typical of young, metabolically active cells); vacuolation (suggesting cellular degeneration or production of secretory products); variation in staining quality; inclusions; small cytoplasmic volume; and variable cytoplasmic margins. Cytoplasmic inclusions include a variety of structures not normally found in the cells, such as small pieces of nuclei (called satellite nuclei); dark irregular structures that may represent degenerate organelles; or phagocytized cells

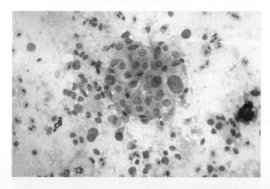

Fig. 19. A ball of epithelial cells with features of malignant neoplasia (anisocytosis, variable N:C ratios, anisokaryosis, multiple nucleoli, possible multinucleation) in the cytology of fluid aspirated from the coelomic cavity of a bird with an ovarian cystadenocarcinoma (Wright's stain, original magnification ×1000).

(cellular cannibalism). A small cytoplasmic volume is indicated by cells with higher than normal N:C ratios, typical of neoplastic cells. An extreme variation in the appearance of the cytoplasmic borders adds to the cellular pleomorphism and is another indication of neoplasia.

The most frequently observed nuclear criteria for malignant neoplasia include nuclear hypertrophy, anisokaryosis, variable N:C ratios, nuclear pleomorphism, abnormal mitoses, abnormal chromatin patterns, multinucleation, irregular nuclear membrane, and abnormal nucleoli. Cells with very large nuclei should be viewed with suspicion of neoplasia. Anisokaryosis (variation in nuclear sizes) is an important feature of neoplasia, especially in cellular aggregates. This also translates into variable N:C ratios. A variable or high N:C ratio in cells that normally have a low ratio is suggestive of malignancy. Nuclear pleomorphism may indicate a rapid mitotic rate or abnormal mitosis. Abnormal mitotic figures (multipolar division) and aggregates of cells exhibiting a high mitotic index are also suggestive of a malignant neoplasm. Cells in cellular aggregates that exhibit varying and unusual nuclear chromatin patterns should also be viewed with suspicion for neoplasia. Irregular, coarse, hyperchromatic chromatin with clear parachromatin spaces is especially suggestive of neoplasia. Multinucleated giant cells, especially cells with an uneven number of nuclei, may indicate asynchronous cell division. Abnormal nucleoli that vary in shape, size, number, and staining quality are additional features of neoplasia. Large (greater than one third the diameter of the nucleus), pleomorphic, or multiple (greater than four) nucleoli should be viewed with suspicion of malignancy. Structural features of malignant neoplasia refer to those features that may aid in the identification of the neoplasm, such as epithelial neoplasia (carcinomas); mesenchymal neoplasia (sarcomas); or discrete cell neoplasia (round cell neoplasms).

Carcinomas originate from epithelial tissue and are characterized by abnormal-appearing epithelial cells. Cellular aggregates occurring in balls, rosettes, or loose cellular groupings are suggestive of adenocarcinomas (malignant neoplasia of glandular epithelium) and the arrangement of these cells may suggest acinar or papillary formation (see Fig. 3). Adenocarcinomas also have the cytologic features of large cytoplasmic secretory vacuoles and giant cell formation.

Mesenchymal neoplasms usually produce poorly cellular samples containing elongated (oval), stellate, or spindle-shaped cells with indistinct cytoplasmic margins that generally do not occur in aggregates (Fig. 20). The cells often appear as individual cells in the cytology sample, and are generally smaller than epithelial cells comparatively. The shape of the nuclei can vary from round, elliptical, and fusiform. An example of a mesenchymal neoplasm is the fibrosarcoma, which is a common sarcoma of mammals and lower vertebrates. Mesenchymal cell neoplasms, such as osteogenic sarcomas, chondromas, and chondrosarcomas, may reveal a heavy eosinophilic background substance (osteoid or chondroid) in the cytologic specimen.

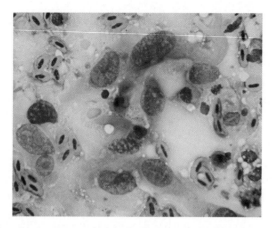

Fig. 20. An aspirate from a mass on the leg of a bird reveals numerous very large mesenchymal cells with features of malignant neoplasia (anisocytosis, anisokaryosis, multiple nucleoli, multinucleation, variable N:C ratios, nuclear pleomorphism, and variable chromatin patterns). The histologic diagnosis was a fibrosarcoma (Wright's stain, original magnification ×1000).

Round cell tumors (discrete cell neoplasia) result from abnormal development of cells that have no normal structural interaction. The cells tend to be round or oval, have distinct cytoplasmic margins, are generally smaller than epithelial cells, and typically exfoliate as individual cells. The nuclei are generally round. Round cell tumors include histiocytomas; mast cell tumors; plasma cell tumors (plasmacytomas); and lymphoma (Figs. 21 and 22) [12].

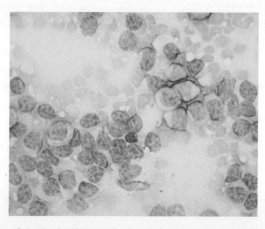

Fig. 21. A lymph node aspiration biopsy of a ferret with lymphoma reveals numerous lymphocytes. Most appear as large lymphocytes, many of which exhibit features of immaturity (Wright's stain, original magnification ×1000).

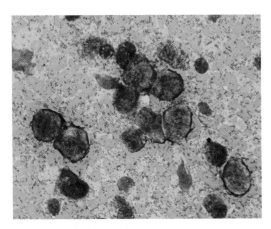

Fig. 22. An aspirate of a cutaneous mass from a tiger cub reveals numerous mast cells indicative of a mast cell tumor (Wright's stain, original magnification ×1000).

## Effusions

In normal animals little if any fluid can be collected from the thoracic and abdominal cavities or coelomic cavity; however, pathologic conditions result in the accumulation of fluid. Fluids accumulate in the body cavities of animals as a result of decreased reabsorption or increased production of normal fluid, an inflammatory process, disruption of vessels or viscus structures, or a neoplastic condition [12]. Before the preparation of fluid samples for cytologic examination macroscopic observations of fluid samples should be made and recorded. These include the source, color, appearance (eg, bloody, mucoid, serosanguinous, watery, milky, or cloudy), and refractometer-determined total solute (protein) concentration of the fluid, and the presence of any odor, clots, or tissue fragments. Quantitative protein concentration of body fluids can be obtained by refractometry or rapid chemical methods, such as biuret [23]. This information facilitates classification of effusions, such as transudates, modified transudates, or exudates.

Transudative effusions are fluids that accumulate in the serous cavities as a result of oncotic pressure changes or other circulatory disturbances (ie, increased hydrostatic vascular pressure). The primary causes of the formation of transudates in domestic mammals include hypoproteinemia (hypoalbuminemia); overhydration; and lymphatic or venous congestion [12]. Cardiac insufficiency, portosystemic shunt, and hepatic cirrhosis and insufficiency are examples of conditions that result in transudates formation. The same causes may result in an increase in coelomic transudative fluid formation in lower vertebrates.

Transudates are typically clear to straw colored grossly, and are characterized by low specific gravity ($<1.017$); low cellularity ($<1000$ cells/µL); and low total protein ($<2.5$ g/dL) [12]. Some authors suggest that the cell counts should be less than 3000 cells/µL [12]. The cells found in transudates are primarily comprised of macrophages, and occasional mesothelial cells, lymphocytes, and nondegenerate neutrophils or heterophils.

Long-standing transudates become modified with the increase in the number of cells or protein content. Modified transudates are often associated with cardiac insufficiency, cardiomyopathy, compression of vascular structures from neoplasia, inflammation or torsion of an organ, and the presence of sterile irritants [24]. Modified transudates grossly resemble transudative effusions, but contain either an increase in cellularity (1000–5000 cells/μL) or protein content (2.5–3 g/dL). The specific gravity is low (1.017–1.025) [12]. The cells found in modified transudates are primarily macrophages and reactive mesothelial cells.

Normal mesothelial cells that have been recently shed into the body cavity appear as flat, polygonal cells with thin, homogenous, weakly basophilic cytoplasm, centrally positioned round or oval nuclei, and occur singly or in clusters of sheets. Mesothelial cells become reactive with irritation of the serous membranes or long-standing effusions. Reactive mesothelial cells are cuboidal to round in shape, are larger in size, exfoliate singly or in clusters, and contain a moderate amount of basophilic cytoplasm compared with nonreactive cells (Fig. 23). The cytoplasm may contain large vacuoles that can push the nucleus to the cell margin creating a signet ring appearance [12]. The nucleus of reactive mesothelial cells tend to be large, round, and often contains coarsely granular chromatin and one to three prominent nucleoli. Many reactive mesothelial cells also have scalloped or villus-like eosinophilic cytoplasmic margins (pink to red fringe). Additional features that might be noted include multinucleation, cytoplasmic vacuolation, and mitotic activity.

Proliferation of mesothelial cells results in the formation of irregular aggregates. These irregular aggregates may exfoliate, appearing in the cytology sample as cellular sheets, balls, or rosettes, which can resemble abnormal cellular clusters seen with some neoplastic disorders (ie, adenocarcinomas).

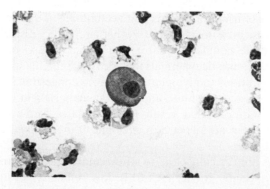

Fig. 23. A reactive mesothelial cell (center) and macrophages in a Cytospin-concentrated fluid sample from the coelomic cavity of a bird (Diff Quick stain, original magnification ×1000).

The uniform appearance of the nuclei in the cell clusters helps to differentiate reactive mesothelial cells from neoplasia.

Exudative effusions are fluids containing increased protein content (>3 g/dL) and cellularity (>5000 cells/μL). Exudative effusions vary in color and turbidity, may have a foul odor, and often clot during sample collection. Fluid samples suggestive of an exudative effusion should be placed into a collection tube containing an anticoagulant (eg, ethylenediaminetetraacetic acid) to prevent clotting of the sample. Exudates typically result from inflammatory processes or chemotactic stimulation within the peritoneal cavity that causes increased capillary permeability. The cellular contents of exudates are primarily inflammatory cells that vary with etiology, host response, and duration of time (Fig. 24).

Hemorrhagic effusions present as red, turbid fluid, and often result from trauma or injury. Peracute hemorrhagic effusions may resemble peripheral blood. Hemorrhagic effusions contain a variable number of erythrocytes, whereas leukocytes and erythrocytes occur in numbers and proportions compatible with peripheral blood (based on cell counts, leukocyte differentials, and the packed cell volume). It is important to differentiate hemorrhagic effusions from peripheral blood contamination of the sample during collection. Observation of the sudden appearance of a red fluid in a clear fluid during sample collection indicates peripheral blood contamination of the sample. The presence of platelets (mammalian sample) or thrombocytes (birds and lower vertebrates) is suggestive of peripheral blood contamination as well, because platelets of mammals and thrombocytes of lower vertebrates disappear quickly in hemorrhagic effusions.

Chronic and resolving hemorrhagic effusions exhibit varying degrees of erythrophagocytosis (erythrophagia). Erythrophagocytosis is indicated by

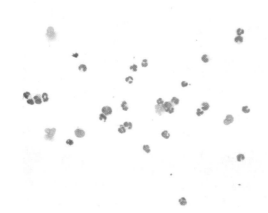

Fig. 24. A direct smear made from a fluid sample obtained by abdominocentesis from a ferret reveals numerous nondegenerate neutrophils and macrophages indicative of an exudative effusion (Wright's stain, original magnification ×1000).

leukocytic (usually macrophagic) phagocytosis of intact erythrocytes, or macrophages containing remnants of erythrocytes, such as red cell fragments and iron pigment (Fig. 25). Iron pigment appears as blue-black to gray pigment in the cytoplasm of macrophages stained with Wright's stain. Hemosiderin crystals (refractile, gold-colored diamond-shaped crystals) also indicate iron from erythrocyte degradation. The presence of iron pigment can be confirmed using Prussian blue stain applied to a Wright's-stained smear.

Chylous effusions of mammals have a "milky" white to pink tinged appearance, and contain variable cell counts and protein content. In general, chylous effusions are classified as either modified transudates or exudates, depending on the degree of chronicity [24]. Long-standing chylous effusions are associated with a mixed population of small mature lymphocytes, vacuolated macrophages, and neutrophils.

Pseudochylous effusions are associated with chronic peritonitis or pleuritis in mammals and differ from true chylous effusions by having higher cholesterol content [24]. A cholesterol-to-triglyceride ratio less than 1 and a triglyceride concentration greater than 100 mg/dL is supportive of a chylous effusion compared with a pseudochylous effusion [25].

Malignant (neoplastic) effusions often result from blood or lymphatic vessel blockage and can have features of modified transudates, hemorrhagic effusions, or exudates and may demonstrate cells with features of malignant neoplasia. Neoplastic cells may be present in the effusion, and their cytologic features may allow the cytologist to classify the malignancy involved (sarcoma, carcinoma, or lymphoid neoplasia). When undifferentiated malignant cells are present in the peritoneal effusion determination of cell origin is very difficult.

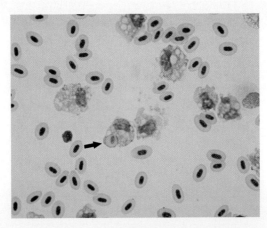

Fig. 25. A smear made from a fluid sample obtained from the coelomic cavity of a bird reveals numerous erythrocytes, macrophages, and erythrophagocytosis indicating a hemorrhagic effusion (*arrow*) (Wright's stain, original magnification ×1000).

## References

[1] Harr K, Alleman A, Dennis P, et al. Morphologic and cytochemical characteristics of blood cells and hematologic and plasma biochemical reference ranges in green iguanas. J Am Vet Med Assoc 2001;218:915–21.
[2] Fox A, Solomon J. Chicken non-lymphoid leukocytes. In: Rose LN, Payne LN, Freeman MB, editors. Avian immunology. Edinburgh: Poultry Science; 1981. p. 135–66.
[3] Dieterien-Lievre F. Birds. In: Rawley AF, Ratcliffe NA, editors. Vertebrate blood cells. Cambridge: Cambridge University Press; 1988. p. 257–336.
[4] Knotkova Z, Doubek J, Knotek Z, et al. Blood cell morphology and plasma biochemistry in Russian tortoises [*Agrionemys horsfieldi*]. Acta Vet (Brno) 2002;71:191–8.
[5] Wigley P, Hulme S, Barrow PA. Phagocytic and oxidative burst activity of chicken thrombocytes to *Salmonella, Escherichia coli* and other bacteria. Avian Pathol 1999;28: 567–72.
[6] Montali R. Comparative pathology of inflammation in the higher vertebrates [reptiles, birds, and mammals]. J Comp Pathol 1988;99:1–26.
[7] Andreasen C, Latimer K. Cytochemical staining characteristics of chicken heterophils and eosinophils. Vet Clin Pathol 1990;19:51–4.
[8] Brooks R, Bounous D, Andreasen C. Functional comparison of avian heterophils with human and canine neutrophils. Comp Haemat Inter 1996;6:153–9.
[9] Harmon B. Avian heterophils in inflammation and disease resistance. Poult Sci 1998;77: 972–7.
[10] Raskin R, Meyer D. Atlas of canine and feline cytology. Philadelphia: WB Saunders; 2001.
[11] Perman V, Alsaker R, Riis R. Cytology of the dog and cat. South Bend (IN): American Animal Hospital Association; 1979.
[12] Baker R, Lumsden J. Cytopathology techniques and interpretation. In: Color atlas of the dog and cat. St Louis: Mosby; 2000.
[13] Klasing K. Avian inflammatory response: mediation by macrophages. Poult Sci 1991;70: 1176–86.
[14] Maxwell M, Robertson G. The avian basophilic leukocyte: a review. World's Poult Sci 1995; 51:307–25.
[15] Maxwell M, Robertson G. The avian heterophil leucocyte: a review. World's Poult Sci 1998; 54:155–78.
[16] Secombes C. The nonspecific immune system: cellular defenses. In: Iwama G, Nakanishi T, editors. The fish immune system: organism, pathogen and environment. San Diego: Academic Press; 1996. p. 63–103.
[17] Montali R, Bush M, Thoen C, et al. Tuberculosis in captive exotic birds. J Am Vet Med Assoc 1976;169:920.
[18] Kennedy-Stoskopf S. Immunology. In: Stoskopf MK, editor. Fish medicine. Philadelphia: WB Saunders; 1993. p. 149–59.
[19] Gyimesi Z, Howerth E. Severe melanomacrophage hyperplasia in a crocodile lizard, *Shinisaurus crocodilurus*: a review of melanomacrophages in ectotherms. J Herp Med Surg 2004; 14:19–23.
[20] Johnson J, Schwiesow T, Ekwall A, et al. Reptilian melanomacrophages function under conditions of hypothermia: observations on phagocytic behavior. Pigment Cell Res 1999; 12:376–82.
[21] Frangioni G, Borgioli G, Bianchi S, et al. Relationships between hepatic melanogenesis and respiratory conditions in the newt, *Titurus carnifex*. J Exp Zool 2000;287: 120–7.
[22] Rebar A. Handbook of veterinary cytology. St. Louis: Ralston Purina; 1978.
[23] George J, O'Neill S. Comparison of refractometer and biuret methods for total protein measurement in body cavity fluids. Vet Clin Pathol 2001;30:16–8.

[24] Shelly S. Body cavity fluids. In: Raskin RE, Meyer DJ, editors. Atlas of canine and feline cytology. Philadelphia: WB Saunders; 2001. p. 187–205.
[25] Waddle J, Giger U. Lipoprotein electrophoresis differentiation of chylous and nonchylous pleural effusions in dogs and cats and its correlation with pleural effusion triglyceride concentration. Vet Clin Pathol 1990;19:80–5.

# Cytologic Diagnosis of Diseases of Rabbits, Guinea Pigs, and Rodents

Michael M. Garner, DVM, DACVP

*Northwest ZooPath, 654 West Main, Monroe, WA 98296, USA*

This article reviews the diseases most amenable to cytologic diagnosis in clinical small mammal practice. Diseases of pet rabbits, guinea pigs, rodents, chinchillas, sugar gliders, and hedgehogs are addressed. The small size of these patients, risk of anesthesia and invasive surgery, and cost factors make small mammals ideal patients for cytologic evaluation when applicable; however, surprisingly few reports exist in the literature, and no other reviews of cytology in these species exist. Much of the data in this article is derived from case submissions to Northwest ZooPath, and disease presentations that seem to be common in this group of animals are emphasized. Diseases of the skin are particularly well represented, especially tumors and inflammatory processes.

## Rabbits

*Neoplasia*

Neoplasia in the rabbit has recently been reviewed [1]. Although reported as uncommon [1], various forms of neoplasia seem to be common in aged pet rabbits, and cytodiagnosis of these tumors is possible based on currently understood cytologic criteria of mammalian tumors in general.

*Basal cell tumors*

Although not well documented in the literature [2], these tumors are particularly common in the author's diagnostic service. Typically, they are solitary benign neoplasms and occur on the trunk or legs. Complete excision of these tumors is usually curative. Rarely, malignant variants are seen and these are invasive tumors with potential for metastasis. These tumors may exfoliate fairly well, and individual or clustered round-to-oval cells are

*E-mail address:* zoopath@aol.com

seen. Typically, these have scant pale blue cytoplasm, minimal anisokaryosis, and small nucleoli. Occasionally, cells have a small amount of cytoplasmic blue-to-black pigment (melanin) (Fig. 1).

## Squamous cell carcinoma

These tumors seem most common in the skin of the ears, face, and feet, and seem to be somewhat common in rabbits [3]. They are invasive neoplasms and have some potential for metastasis. These tumors often are markedly inflamed, and inflammation can mask the neoplastic component in cytologic preparations. The cytomorphologic features of these tumors are identical to those of other mammals [4], taking into account that the inflammatory component contains heterophils rather than neutrophils.

## Trichoepitheliomas

These tumors are essentially basal tumors arising from the adnexa that have variable follicular and squamous differentiation and may be associated with production of keratin. A vast list of synonyms exist for these tumors, including trichofolliculoma, keratoacanthoma, tricholemmoma, and basosquamous tumor. Most of these are benign, and they have similar clinical appearance and distribution as basal cell tumors, although on cut surface they may have a laminated core of keratin or a cystic appearance. Cytologic appearance may be similar to benign basal cell tumor or may resemble the similar tumors seen in guinea pigs, with some keratinaceous debris or squamous cells present (Figs. 2 and 3).

## Mammary gland tumors

Tumors and cysts of the mammary glands are occasionally seen in laboratory and pet rabbits, and most of the tumors are adenocarcinomas. These tumors are sometimes found concurrently in rabbits that have uterine

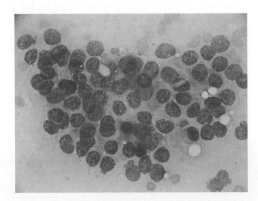

Fig. 1. Aspirate of skin mass, basal cell tumor, rabbit. Note clusters of round to slightly elongated cells with scant cytoplasm, uniform small nuclei, and small nucleoli (Wright-Giemsa).

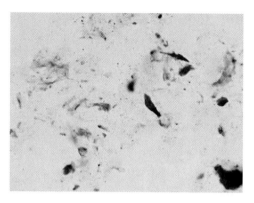

Fig. 2. Aspirate of cystic trichoepithelioma from dorsal rump, guinea pig. Note that sample is hypocellular, with shards of keratin and some cellular debris (Wright-Giemsa).

adenocarcinomas [5,6]. Because these tumors can be cystic or inflamed, the cytologic appearance can be highly variable, and does not always correlate well with histologic findings. Individual or small clusters of variably differentiated epithelial cells may be admixed with proteinaceous fluid matrix, blood, and mixed inflammatory cells (Fig. 4).

*Lymphoma*

Lymphoma is one of the most common malignant neoplasms of rabbits and is the most common malignancy of young rabbits [3]. Typically, this tumor has a multicentric distribution that may involve tissues accessible for cytologic evaluation, such as lymph node, skin, oral cavity, and eye. Cellular morphology varies depending on the tumor grade; however, most lymphoid malignancies in rabbits are high grade, and the cells have considerable

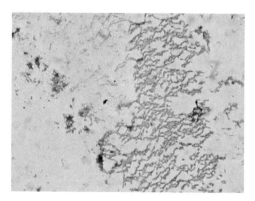

Fig. 3. Aspirate of cystic trichoepithelioma from dorsal rump, guinea pig. Note that sample is hypocellular and comprised entirely of extracellular poorly staining granular material consistent with sebum (Wright-Giemsa).

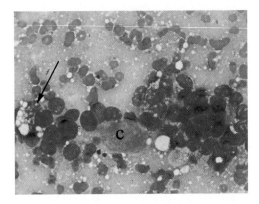

Fig. 4. Aspirate of mammary gland carcinoma, rabbit. Note cluster of epithelial cells with variability in nuclear size and shape, and basophilic mildly vacuolated cytoplasm. A vacuolated macrophage is present (*arrow*), and there is a proteinaceous matrix suggesting cystic fluid. The eosinophilic amorphous material (c) may represent corpora amylacea, a form of inspissated mammary secretion (Wright-Giemsa).

anaplasia, including increased cytoplasmic basophilia, large nuclei with multiple nucleoli, high nuclear/cytoplasmic ratio, and mitoses (Fig. 5).

*Soft tissue sarcomas*

Although not well documented in the literature, soft tissue sarcomas involving the skin and underlying muscle are common in middle-aged to old rabbits in the author's practice. These tumors often are not well differentiated, and the cell of origin can be difficult or impossible to determine from cytologic or histologic examination. Fibrosarcoma,

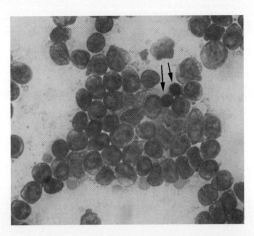

Fig. 5. Aspirate of submandibular lymph node, lymphoma, guinea pig. Note population of lymphoblasts that predominates. Few well-differentiated small lymphocytes are also in the field for comparison (*arrows*) (Wright-Giemsa).

rhabdomyosarcoma, leiomyosarcoma, and histiocytic sarcoma are probably the most common. Typically, these tumors are highly invasive and difficult completely to excise, although rate of metastasis may be slow. These tumors have variable and sometimes poor cellular exfoliation, and the cells vary from round to spindloid. Anaplasia may be striking in these tumors, with marked anisokaryosis, multiple nucleoli, and binucleated or multinucleated cells (Fig. 6).

*Lipoma*

Lipoma is a somewhat common benign subcutaneous tumor in rabbits, usually involving the trunk. As in other species, it can be difficult to distinguish lipoma from aspiration of normal fat, although clinical appearance of the tumor and history in conjunction with the cytologic findings are often helpful in making this distinction. Typically, these tumors exfoliate poorly and only rare individual or small clusters of normal-appearing lipocytes may be found (Fig. 7).

*Uterine adenocarcinoma*

The most common neoplastic process involving the abdominal cavity of female rabbits is uterine adenocarcinoma [3]. Affected rabbits may have abdominal distention and a palpable mass affect. Cytologic diagnosis should be made conservatively, because there is considerable hyperplastic or dysplastic change in the uterus of rabbits with cystic endometrial hyperplasia or adenomyosis, and these nonneoplastic conditions could be confused with neoplastic transformation of the endometrial glands. Uterine adenocarcinoma can metastasize and may be associated with seeding of the abdominal cavity (carcinomatosis), and it is possible that aspiration of a uterine tumor could lead to seeding of the abdominal cavity (Fig. 8). Exfoliation

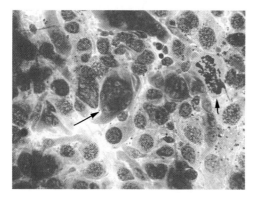

Fig. 6. Aspirate of giant cell sarcoma, leg region, rabbit. Note high degree of cellular anaplasia characterized by basophilia, pleomorphism with marked anisokaryosis, bizarre mitotic figure (*short arrow*), and frequent multinucleated cells (*long arrow*) (Wright-Giemsa).

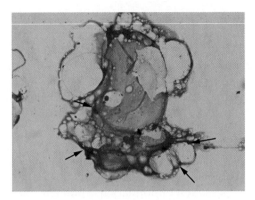

Fig. 7. Aspirate of lipoma from leg, guinea pig. Note cluster of variably sized pale blue adipocytes with large spherical vacuolated cytoplasm and displaced, compressed nuclei (*arrows*) (Modified Wright-Giemsa).

of neoplastic cells can be seen in abdominal effusions associated with this tumor. Also, cytologic differentiation of this tumor from other less common intra-abdominal tumors, such as pancreatic or intestinal adenocarcinoma, biliary adenocarcinoma, or embryonal nephroma, may be difficult.

*Testicular tumors*

Testicular tumors are occasionally observed in rabbits, and the most common on file at Northwest ZooPath is the benign interstitial or Leydig cell tumor. These tumors are comprised of sheets of large polygonal cells with moderate to large amounts of lightly basophilic and occasionally foamy cytoplasm and small ovoid nuclei with single small nucleoli (Fig. 9). Rarely seminoma, Sertoli cell tumor, or testicular teratoma may also occur in rabbits.

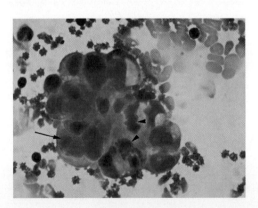

Fig. 8. Abdominal fluid aspirate, uterine adenocarcinoma, rabbit. Note cluster of neoplastic epithelial cells with marked anisokaryosis, large nucleoli (*arrow*), and mitotic figures (*arrowheads*) (Wright-Giemsa).

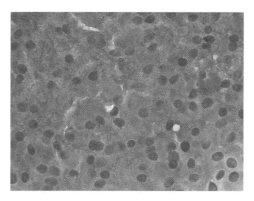

Fig. 9. Aspirate of testicular interstitial cell tumor (Leydig cell tumor), rabbit. Note sheets of polygonal cells with abundant granular cytoplasm and small nuclei (Modified Wright-Giemsa).

## Thymoma

With the exception of metastatic tumors and lymphoma, thymoma is probably the most common intrathoracic neoplasm in rabbits [6]. These tumors generally are benign but can impinge on lung capacity and are occasionally associated with paraneoplastic syndromes, such as hypercalcemia or dermatopathies. The tumors can be lymphoid, epithelioid, or mixed and cytologic examination reflects these variants (Fig. 10).

## Nonneoplastic conditions

### Polyps

Polyps involving the rectum, vulva, or penile sheath are common in rabbits. Although the etiology of this condition has not been determined, the

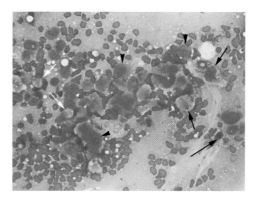

Fig. 10. Aspirate of thymoma, rabbit. Note clusters of epithelioid cells with large nuclei and variable amounts of basophilic cytoplasm (*arrowheads*). Erythrocytes, scattered macrophages (*short arrows*), a lymphocyte (*long arrow*), and some heterophils (*white arrows*) are also present (Wright-Giemsa).

distribution and morphology of these lesions and their similarity to viral-induced lesions in other mammals suggests that they may have a viral etiology. Typically, these lesions are comprised of well-differentiated but hyperplastic squamous epithelium with hyperkeratosis. The epithelial component is generally supported by an inflamed fibrous stroma. These lesions often have a secondary bacterial or traumatic component, leading to erosion or crusting of the surface. Cytologic examination reveals a mixture of well-differentiated squamous epithelium, some parabasal cells, and varying degrees of mixed inflammation and bacteria (Fig. 11).

*Squamous papilloma*

These lesions are occasionally seen in wild and domestic rabbits, primarily involving the face or shoulder regions, and are caused by papillomaviruses. The Shope papillomavirus causes benign disease in wild rabbits, but there is a high incidence of transformation to squamous cell carcinoma in domestic rabbits infected with this virus. Oral papillomatosis of domestic rabbits is caused by oral papillomavirus, and these lesions typically regress over time [7]. Rabbit papillomas have morphology typical of viral papilloma, with marked papilliform hyperplasia of the epidermis, a prominent granular cell layer, and occasional koilocytes in the granular layer. Aspiration is less productive than impression smears. Squamous cells predominate in cytologic preparations, although few parabasal cells and perhaps few atypical cells with anisokaryosis may be seen (Fig. 12).

*Nodular dermal fibrosis*

This condition occurs in the skin of rabbits, primarily over the back and on the flanks, and presents as small barely visible or palpable masses in the dermis. These are comprised of broad bands of dermal collagen delineated

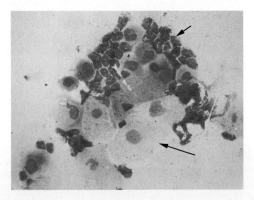

Fig. 11. Impression smear from rectal polyp, rabbit. Note well-differentiated squamous epithelial cells (*long arrow*) and few parabasal cells admixed with clusters of degenerative neutrophils (*short arrow*) (Modified Wright-Giemsa).

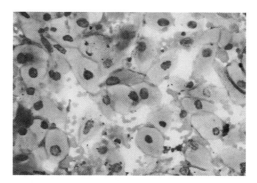

Fig. 12. Impression smear of squamous papilloma from skin of face, rabbit. Note uniform population of well-differentiated squamous epithelial cells and some erythrocytes (Wright-Giemsa).

by low numbers of well-differentiated fibrocytes. The epidermis may be mildly hyperplastic. Inflammation and adnexal hyperplasia are rarely seen. The etiology of this condition has not been determined but these may represent developmental hamartomas or foci of previous trauma. These lesions do not exfoliate well, and aspirates usually contain scant blood and possibly a few bland spindle cells (fibrocytes).

*Abscesses*

Abscesses are particularly common in rabbits [8], and cytologic examination can be useful in establishing a diagnosis. They occur most commonly in the facial soft tissues as manifestations of oral trauma or tooth root problems. Abscesses are also common in the abdominal cavity, presumably arising in the mesenteric lymph nodes, following bacterial embolization from the gut. *Pasteurella multocida*, *Escherichia coli*, and *Staphylococcus* sp are the most common isolates. Cutaneous abscesses often are well encapsulated and may be amenable to surgical excision. These lesions typically exfoliate large numbers of heterophils and macrophages with fewer plasma cells, lymphocytes, and blood. Cellular debris indicative of necrosis, mineral, and intracellular or extracellular bacteria may be seen (Fig. 13). Rabbit abscesses sometimes have a core of eosinophilic debris resembling Splendore-Hoeppli material, especially with staphylococcal infection (Fig. 14). Facial abscesses often are associated with concurrent suppurative nasal exudate that has a similar cytologic appearance and needs to be differentiated from chronic rhinitis of pasteurellosis and treponemiasis.

*Cutaneous treponemiasis (rabbit syphilis)*

Rabbit syphilis is a relatively common cutaneous infectious disease. Affected animals have crusts and plaque-like lesions involving the skin and oral mucous membranes of the face and genitalia, although lesions are

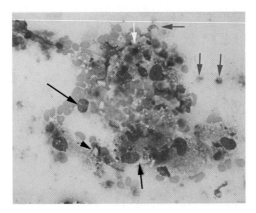

Fig. 13. Aspirate from a large intra-abdominal abscess, rabbit. Note mixed population of heterophils (*long arrow*), macrophages (*short arrow*), and erythrocytes, with some extracellular mineral (*arrowhead*). Much cellular necrosis is present (*red arrows*), and one macrophage contains phagocytized cellular debris (*white arrow*) (Wright-Giemsa).

sometimes also seen in the dorsum of the back. The primary lesion is epidermal hyperplasia and variable hyperkeratosis or parakeratosis. Moderately severe inflammation is present in the dermis of these lesions, primarily heterophils (Fig. 15). The bacteria that cause the lesion are slender filamentous organisms that are present between the epithelial cells, usually in deeper regions of the epidermis, and in the infundibular regions of the follicular sheath. These organisms are difficult to detect without the aid of Warthin-Starry or Steiner silver stains, which can be applied to cytologic preparations (Fig. 16). Biopsy often is needed to confirm this diagnosis.

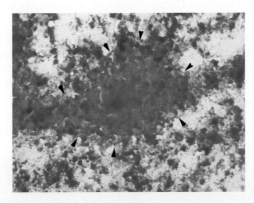

Fig. 14. Aspirate from leg abscess, rabbit. Note accumulation of amorphous pink granular material that may represent Splendora-Hoeppli reaction (*arrowheads*), surrounded by degenerative heterophils (Modified Wright-Giemsa).

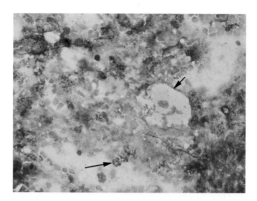

Fig. 15. Scraping from a focus of facial treponemiasis, rabbit. Note squamous epithelium (*short arrow*) and numerous viable or degenerative heterophils (*long arrow*). Note that spirochetes are not visible (Wright-Giemsa).

## Guinea pigs

### Neoplastic conditions

Current literature suggests that spontaneous neoplasia of guinea pigs (cavies) is somewhat uncommon [9]; however, various neoplastic processes constitute the bulk of cavy submissions at the author's practices, and some forms of neoplasia are perhaps understated in the literature.

### Adipocyte tumors

These tumors are very common in guinea pigs. They typically occur on the ventral abdomen and less commonly on other parts of the trunk. They may be solitary or multicentric, and are soft and freely moveable.

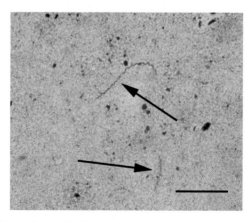

Fig. 16. Scraping from a focus of facial treponemiasis, rabbit. Note few argyrophilic, extracellular spirochetes (*arrows*). (Warthin Starry). Bar = 10 μm.

Lipoma is most common, and aspiration yields small clusters of well-differentiated adipocytes (see Fig. 7). Liposarcoma is occasionally seen in the skin and subcutis of cavies, and these cells have plump, anisokaryotic nuclei with variable cytoplasmic vacuolar change (Fig. 17).

## Trichoepithelioma

These tumors are common in guinea pigs [9]. They are especially common in males over the lumbar region, where scent glands are numerous. Variation in morphology may occur, but usually these tumors are comprised of follicular formations of basal cells that differentiate to sebaceous or squamous epithelium, these structures often oriented concentrically around a central cavity containing laminated keratin, inflammatory cells, or debris. Cytologic appearance varies according to the site aspirated within the tumor. Because the cystic portion of the tumor may predominate, many aspirates contain only keratin or sebum, perhaps with few inflammatory cells or blood (see Figs. 2 and 3). Occasionally, a basal cell component or keratinocytes may be represented in the sample. Excision of these benign neoplasms generally is curative. They become a problem if they ulcerate, and are prone to fly strike.

## Mammary tumors

Proliferative lesions of the mammary glands are common in female and male guinea pigs [9]. Lesions include glandular or ductular cyst, adenoma or cystadenoma, and adenocarcinoma. The prevalence of these three lesions is about equal at the author's practices. There may be a continuum in the development of these lesions. As with rabbits, cytologic appearance of the lesions varies depending on the stage of development. Cysts or cystic neoplasms may exfoliate poorly and be comprised only of proteinaceous fluid

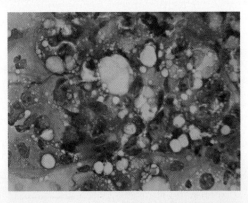

Fig. 17. Aspirate of liposarcoma from body wall, guinea pig. Note cluster of anaplastic adipocytes with large plump, variably sized nuclei, some with multiple nucleoli. Cytoplasmic vacuolization is quite variable and there is less or no nuclear displacement, compared with lipoma in previous image (Wright-Giemsa).

matrix with few inflammatory cells. Tumors with a more solid tumor component exfoliate clusters of atypical epithelial cells (see Fig. 4). There may be much overlap in the cytologic appearance, and biopsy may be needed fully to characterize these lesions. Reports of guinea pig mammary gland fibroadenoma exist, but in the author's experience this is a misnomer. The tumors in cavies do not have a significant fibrous component, and do not resemble the classic mammary gland fibroadenoma seen in rats.

*Lymphoma*

Lymphoid malignancies (lymphoma and lymphoblastic or "cavian" leukemia) are perhaps the most common form of malignancy in guinea pigs [9]. As with other species, lymphoma is generally multicentric and may involve skin, facial structures, lymph nodes, and viscera. These tumors may have variable cell morphology, although high-grade blast cell forms are most common (see Fig. 5).

*Bronchogenic papillary adenoma*

Bronchogenic tumors are common in guinea pigs [9], are almost always benign, and can be multicentric and involve multiple lobes. Because the tumors can be radiographically apparent, they may be confused for metastatic cancer, abscesses, or granulomas. Generally, these tumors have benign cytomorphology, so accurate aspiration may be helpful in establishing a diagnosis.

*Thyroid tumors*

Few reports exist of thyroid neoplasia in guinea pigs [10,11], although these are one of the most common tumors seen in this species in the author's practice. These tumors usually are unilateral, and may be benign or malignant. The malignant tumors are generally well differentiated and may contain bone. The cytologic appearance is clustered round-to-cuboidal cells with varying degrees of atypia depending on benign or malignant variants. Aspirates often contain much blood if the tumors are cystic, or variable proteinaceous matrix if the tumors are producing colloid (Fig. 18). The occurrence of pigment in the cytoplasm of neoplastic epithelial cells sometimes seen cytologically in canine and feline thyroid tumors is unusual in the author's experience.

*Leiomyoma of uterus*

Benign smooth muscle tumor is the most common uterine tumor in guinea pigs seen at the author's practice. These tumors can achieve large size and may be associated with abdominal enlargement, vulvar effusion, and clinical morbidity. They may be palpable and can be aspirated, if done with care. Typically, these tumors exfoliate low to moderate numbers of cells. Cells generally are spindloid with moderate amounts of cytoplasm

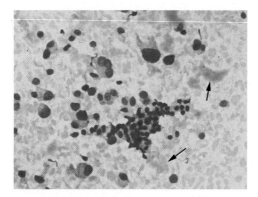

Fig. 18. Aspirate of thyroid adenocarcinoma, guinea pig. Note clustered and individual atypical epithelial cells that have variable amounts of blue cytoplasm, moderate to marked anisokaryosis, and binucleated or multinucleated cells. Note the basophilic extracellular material (*arrows*), which may represent colloid (Wright-Giemsa).

and central elongated nuclei that have blunt poles. Anisokaryosis is usually mild.

*Transitional cell carcinoma*

Old cavies may have a relatively high prevalence of transitional cell carcinoma of the urinary bladder, based on submissions to the author's practice. These tumors typically exfoliate large numbers of cells into the urine, and these may have a high degree of cellular anaplasia (Fig. 19) or cytoplasmic vacuolar change (Fig. 20). Because these tumors sometimes are ulcerated or obstructive, there may be concurrent inflammation, hemorrhage, or mineralized debris in the cytologic preparations.

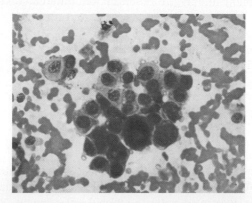

Fig. 19. Urine sediment, transitional cell carcinoma, guinea pig. Note cluster of highly anaplastic epithelial cells with considerable anisocytosis and anisokaryosis, large nucleoli, and binucleated or multinucleated cells. Also note the large numbers of erythrocytes consistent with hemorrhage (Wright-Giemsa).

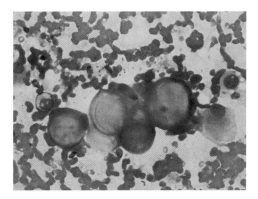

Fig. 20. Urine sediment, transitional cell carcinoma, guinea pig. Note cytoplasmic vacuolar change with peripheral displacement of nuclei (Wright-Giemsa).

## Nonneoplastic conditions

### Lymphadenitis

Cervical lymphadenitis is common in cavies. In young cavies the condition is often associated with *Streptococcus zooepidemicus* or *Streptococcus pneumoniae*, and is a contagious disease [12]. In older cavies, cervical lymphadenitis is more often associated with migrating foreign body or oral trauma. Cytologic appearance of this lesion is that of an abscess, and because by the time the condition is clinically apparent the node is effaced, few if any lymphoid cells may be present (Fig. 21). Bacteria are only rarely identified in these lesions in the author's experience.

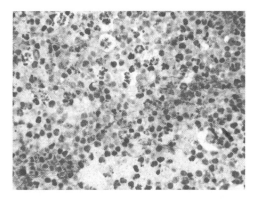

Fig. 21. Aspirate of submandibular lymph node, cervical lymphadenitis, guinea pig. Note large number of viable and degenerative neutrophils admixed with small lymphocytes (Wright-Giemsa).

*Hepatic lipidosis*

Diffuse fatty liver is common in guinea pigs and can be associated with significant clinical morbidity and mortality. The condition may be seen with anorexia, obesity, malnutrition, endocrine or metabolic derangements, hypoxia, and hepatotoxins. It seems particularly common in periparturient cavies with the metabolic-nutritional form of pregnancy toxemia [13]. Although the condition is potentially reversible, administration of good supportive care and adequate nutrition during the interim can be problematic, and mortality may be high in these patients. Accurate hepatic aspirate is diagnostic, and reveals marked lipid vacuolization in hepatocytes (see Fig. 9 in the article on hedgehogs elsewhere in this issue).

*Ovarian cysts*

Ovarian cysts are very common in adult cavies and usually are bilateral [14]. They are sometimes functional and associated with hormonal imbalances that manifest as reduced reproductive performance, cystic endometrial hyperplasia, mucometra, endometritis, fibroleiomyomas, and alopecia. They can achieve large size and may be palpable. Aspirates usually have very low or no cell yield and are comprised predominantly of serous to proteinaceous fluid. Aspiration should be performed with care, because rupture or continued leakage may lead to sterile peritonitis.

*Kurloff's bodies*

In guinea pigs, natural killer cells of the lymphoid lineage contain large intracytoplasmic pink-to-purple inclusions referred to as "Kurloff's bodies," and these cells may be seen in peripheral blood films or in cytologic preparations of spleen or bone marrow. These mucopolysaccharide inclusions are unique to cavies, but are of unknown function. They sometimes are mistaken for viral or parasitic inclusions (Fig. 22) [15].

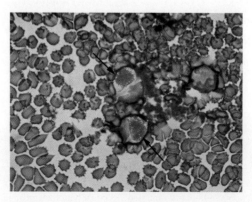

Fig. 22. Blood film, guinea pig. Note large lymphocytes with cytoplasmic inclusions (*arrows*) typical of Kurloff's bodies (Wright-Giemsa).

## Rats

Rats are popular pets. They are affectionate animals that adapt well to handling, are easy to feed and clean, and are relatively malodorous. Unfortunately, they have a short lifespan and a variety of spontaneous diseases. Some of these diseases are amendable to cytologic interpretation and are discussed and illustrated next.

*Neoplastic conditions*

*Mammary gland tumors*

The most common mammary tumor of rats is fibroadenoma [16]. These tumors can be solitary or multicentric along the mammary chain and also commonly occur in seemingly ectopic mammary tissue near the tail, over the back, and around the neck. These tumors can achieve large size if not excised, and can impair mobility. Large tumors often ulcerate because of external trauma and may become secondarily infected. These tumors exfoliate poorly but may have a cytologic appearance that includes scattered bland fibroblasts and occasional individual or clustered glandular epithelial cells that have little atypia. In ulcerated, infected tumors the neoplastic component may be masked by inflammatory cells.

Although less commonly discussed in pet rat medicine, these tumors are somewhat common in the author's practice. They are aggressive malignancies with considerable potential for local tissue invasion and metastasis. Typically, these are tubulopapillary neoplasms that may be cystic and inflamed, and have the same cytologic criteria and ambiguities that apply to glandular mammary tumors in other species.

*Lipoma*

These tumors are common in rats and have the same biologic behavior and cytologic characteristics as in other species.

*Mesenchymal tumors of the skin and peripheral soft tissues*

Spindle cell tumors are common in middle-age to older rats, particularly fibrosarcoma. Histiocytic sarcoma and solid manifestations of monocytic leukemia are also occasionally encountered.

*Zymbal's gland tumor*

Rats are prone to development of a tumor that arises in the glands of the external ear canal called a Zymbal's gland tumor [17]. These tumors are invasive carcinomas that have variable glandular differentiation, and occasional squamous differentiation. These tumors frequently are inflamed and may be scirrhous, and these features may be reflected in the cytologic appearance. Rate of metastasis seems slow.

*Interstitial cell tumor*

These are perhaps the most common of the rat testicular neoplasms. They generally are unilateral, and considered benign neoplasms. Their cytomorphologic appearance is similar to that of other mammalian species.

*Lymphoma and other hematopoietic tumors*

Hematopoietic neoplasms are particularly common in rats, especially solid lymphoid malignancies and large granular lymphocytic leukemia [18]. These tumors are typically multicentric, involve multiple organs, and warrant a poor prognosis.

*Nonneoplastic conditions*

*Bronchopneumonia (chronic murine pneumonia)*

Bronchopneumonia is particularly common in middle-age to older pet store and feeder rats, and is associated with a variety of infectious agents, including *Mycoplasma pulmonis*, cilia-associated respiratory bacillus, and a variety of bacterial agents [9]. This disease responds at least partially to antibiotic therapy but tends to relapse. Occasionally, affected rats may have a nasal or thoracic effusion or an intrathoracic abscess large enough to aspirate, and cytologic examination may reveal suppurative inflammation, with or without bacteria.

*Vaginitis and pyometra*

These conditions seem to be particularly common in pet rats in the author's practice, although perhaps are underdocumented in the literature. Vaginitis and open pyometra may be associated with a visible vaginal discharge. Cytologic evaluation of the effusion reveals neutrophils and mucus, sometimes admixed with extracellular or intracellular bacteria (Fig. 23). Care should be taken in considering percutaneous aspiration of a closed

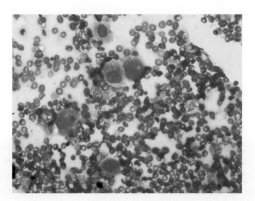

Fig. 23. Smear of vaginal effusion, rat. Note large amount of blood, neutrophils, and macrophages (Wright-Giemsa).

pyometra, because there may be some risk of developing peritonitis from the procedure.

*Pyoderma*

Pyoderma is a common malady of pet rats and mice [19]. These animals typically are pruritic and may have ulcerated skin, especially over the shoulders or back. Imprints of the lesions reveal a suppurative exudate admixed with bacteria, usually cocci resembling *Staphylococcus* sp (Fig. 24). Rarely, mites or conspecific trauma may also be involved in the pathogenesis of these lesions.

## Mice

Although mice are popular pets, the author's diagnostic service has few cytology submissions on file. Cytologic evaluation seems like a viable alternative to the more costly and invasive biopsy procedures performed on these small animals, and it is surprising that more lesions are not examined cytologically for this group.

*Neoplasia*

Tumors of the skin or associated soft tissues are uncommonly encountered in pet mice. Those seen most frequently in the author's diagnostic service are thyroid adenoma or adenocarcinoma, lymphoma, granular lymphocytic leukemia (Fig. 25), and squamous cell carcinoma. Less commonly seen tumors include myelogenous leukemia, hepatocellular carcinoma, renal cell carcinoma, and thymoma.

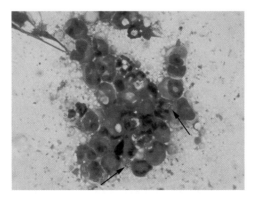

Fig. 24. Swab of ulcerated skin lesion, pyoderma, rat. Note predominance of neutrophils and few macrophages. Few neutrophils contain phagocytized cocci (*arrows*) (Wright-Giemsa).

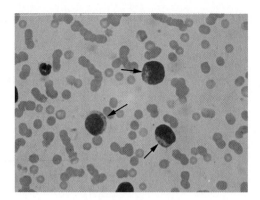

Fig. 25. Blood film, large granular lymphocytic leukemia, mouse. Note large lymphoid cells with basophilic cytoplasm that contain few large pink intracytoplasmic granules (*arrows*) (Wright-Giemsa).

*Bacterial infections*

The most common skin condition in mice is staphylococcal dermatitis (pyoderma), and is sometimes associated with concurrent ectoparasitism or conspecific aggression [20]. Imprints of affected skin yield squamous epithelium, mixed inflammation with predominance of neutrophils, and occasional extracellular or intracellular cocci. A number of other bacterial conditions involving the skin or soft tissues of the trunk, face, and legs may be seen, especially abscesses associated with bite wounds, in which a number of different gram-positive or gram-negative bacteria may be isolated or observed microscopically.

## Hamsters

Hamsters have a few somewhat unique neoplastic processes primarily involving the skin that may be diagnosed cytologically.

*Lymphoma and malignant plasma cell tumor*

These tumors are common in hamsters, and as in other species, typically are multicentric malignances that may involve skin, lymph nodes, and viscera [9]. Cytologic characteristics of lymphoma in hamsters generally are as seen in other mammals. A plasmacytoid form of lymphoma, also sometimes referred to as "malignant plasma cell tumor," seems particularly common in hamsters. Cytologically, these tumor cells have plasmacytoid characteristics that include eccentric round-to-ovoid nuclei; stippled chromatin; basophilic cytoplasm; occasional perinuclear vacuole (Golgi zone); and sometimes few binucleated or multinucleated cells (Fig. 26). Lymphoid,

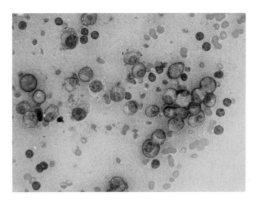

Fig. 26. Aspirate of skin mass, plasma cell tumor, hamster. Note plasmacytoid round cells with slightly eccentric nuclei, blue cytoplasm, and small perinuclear clear zone (Golgi zone) (Wright-Giemsa).

myeloid, and plasmacytoid malignancies of hamsters may occur spontaneously, but also have been associated with papovavirus infection.

*Melanoma*

Cutaneous melanoma is occasionally seen in hamsters [21]. These tumors exfoliate epithelioid or histiocytoid cells that have varying amounts of cytoplasmic dark granular material representing melanin pigment production. Cellular anaplasia is highly variable, as is degree of cellular exfoliation (Fig. 27). Also, the pigment of melanoma cells can be confused with hemosiderin of chronic hemorrhage; poorly exfoliative and concurrently inflamed melanoma can be difficult to diagnose cytologically.

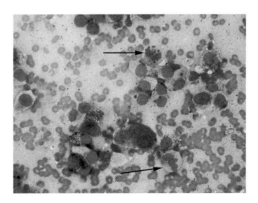

Fig. 27. Aspirate of skin mass, melanoma, hamster. Note round cells with considerable anaplasia, including anisokaryosis, binucleated cells, and mitotic figures (*arrows*). Most of the cells contain cytoplasmic black pigment (melanin) (Wright-Giemsa).

## Trichoepithelioma and papilloma

These lesions may be associated with a papovavirus, and this condition has also been positively correlated to the development of lymphoid malignancies [22]. Hamsters are prone to developing papillomas, especially involving the ear pinna and external ear canal. Cytomorphology is similar to that of squamous papilloma or trichoepithelioma in other species; the diagnosis of trichoepithelioma or papilloma should warrant a search for concurrent or future lymphoma in the patient.

## Mammary gland adenocarcinoma

These tumors may be common [21] and are the most common tumors on file for hamsters at the author's practice. All cases on file are malignant, although follow-up information regarding metastatic behavior is scant. These tumors exfoliate in a manner typical of mammary gland adenocarcinoma in other species.

## Miscellaneous tumors

Miscellaneous tumors that occasionally occur in hamsters and may be encountered cytologically include adrenal cortical adenoma or carcinoma; proliferative lesions of the ventral abdominal marking gland (hyperplasia, adenoma, adenocarcinoma, or squamous cell carcinoma); and spindle cell tumors involving the skin.

# Mongolian gerbils

The incidence of spontaneous neoplasia in gerbils is reported as high by some authors [21] or low by others [9]. All seem to agree that tumors occurring in the ovary are the most common form of neoplasia in this species. Granulosa cell tumors, lutein tumors, and thecal cell tumors have been described. Proliferative lesions of the ventral abdominal marking gland, including hyperplasia, adenoma, adenocarcinoma, and squamous cell carcinoma, are also common and constitute the most common cytologic diagnosis at the author's practice regarding this species.

# Prairie dogs

Prairie dogs are popular exhibit animals at a number of United States and international zoos. They have also become popular pets in the United States, but recently have been linked to the spread of monkey pox to humans, and this may adversely impact the pet trade in this species. There are a few diseases that are somewhat common in prairie dogs and may be encountered cytologically.

## Hepatocellular carcinoma

Hepatocellular carcinoma is perhaps the most common tumor seen in prairie dogs [23]. A similar tumor is seen in related woodchucks and ground squirrels and associated with hepadenavirus infection. Viral infection also has been proposed for the prairie dog tumors but has not yet been proved. There are at least four histologic patterns of these tumors but cytologically these differences may not be apparent. These tumors are comprised of hepatoid cells that have variable degrees of anaplasia (Fig. 28). Hemodilution or predominance of necrotic cell debris can make these tumors difficult to diagnose cytologically, although cellular exfoliation may be good in solid viable portions of the tumor. Well-differentiated hepatocellular carcinoma can be difficult to distinguish from hepatocellular adenoma (hepatoma) in cytologic preparations.

## Odontoma

Odontomas are particularly common in prairie dogs [24] especially in the maxillary arcade, although it is questionable that these are true tumors, and may be reactive processes associated with dental trauma. The tumors are composed of well-differentiated but mildly disorganized tooth structures, and in some cases cannot be distinguished from normal teeth. These tumors exfoliated poorly or not at all, but frequently occlude the nasal passage and may be associated with secondary bacterial rhinitis. Odontoma should be considered for cytologic preparations of suppurative exudate from the nasal cavity.

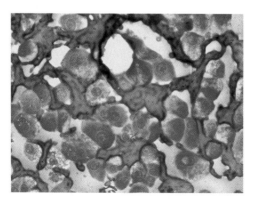

Fig. 28. Aspirate of hepatocellular carcinoma, prairie dog. Note hepatoid cells with moderate amount of pink cytoplasm, round nuclei with prominent nucleoli, mild anisokaryosis, and few binucleated cells (Wright-Giemsa).

*Facial abscesses and necrotic stomatitis*

Prairie dogs seem prone to developing facial abscesses along the dental arcades, and these are likely caused by oral trauma or tracts from infected tooth roots. *Actinomyces* sp and various mixed anerobic and aerobic bacteria may be isolated from these lesions. Cytologic presentation is that of an abscess or suppurative exudate, and bacteria may or may not be present. In some cases, these animals also have necrotizing stomatitis or gingivitis. The author has seen at least one case of herpesvirus associated with ulcerative gingivitis in prairie dogs.

*Hepatic lipidosis*

One of the most common problems facing prairie dog owners and clinicians is control of obesity in prairie dogs. Overfeeding of rich foods and reduced access to exercise in these otherwise very active animals likely account for this problem in captivity. Obese prairie dogs may develop hepatic lipidosis, which may become severe enough to cause clinical hepatopathy [25]. Cytologic diagnosis can be achieved as with other animals prone to this condition.

*Miscellaneous tumors*

The author has a list of miscellaneous tumors on file from prairie dogs that may be amenable to cytologic diagnosis. These include thyroid adenocarcinoma, mammary gland adenocarcinoma, uterine leiomyosarcoma, lymphoma, lipoma, and bronchiolo-alveolar adenocarcinoma.

**Chinchillas**

Significant disease processes in the chinchilla that may be applicable to cytologic diagnosis include cutaneous and facial or tooth root abscesses, lymphoma, cutaneous lipoma, hepatic lipidosis, and cavitary effusions associated with degenerative cardiac disease. Relative to most other rodents, the incidence of neoplasia in chinchillas, a long-lived species, is surprisingly low [21].

**Sugar gliders**

Sugar gliders are relatively new to the pet scene, and the author has few cytologic samples on file from this species. They do not seem to be prone to pyoderma, although traumatic skin disease seems somewhat common and often has a bacterial component. Neoplasia seems uncommon. Those tumors most frequently seen in the author's practice include mammary gland adenocarcinoma, lymphoma, squamous cell carcinoma, and cutaneous melanoma. These animals also seem prone to biliary cysts and cystadenoma, hepatocellular tumors, and hepatic iron storage.

# References

[1] Heatley JJ, Smith AN. Spontaneous neoplasms of lagomorphs. Vet Clin Exot Anim 2004;7: 561–77.

[2] Li X, Schlafer D. A spontaneous skin basal cell tumor in a black French mini-lop rabbit. Lab Anim Sci 1992;42:94–5.

[3] Weisbroth S. Neoplastic disease. In: Manning P, Ringler D, Newcomber C, editors. The biology of the laboratory rabbit. 2nd edition. New York: Academic Press; 1994. p. 259–92.

[4] Tyler RD, Cowell RL, Meinkoth JH. Cutaneous and subcutaneous lesions: masses, cysts, ulcers and fistulous tracts. In: Cowell RL, Tyler RD, editors. Diagnostic cytology of the dog and cat. Goleta: American Veterinary Publications; 1989. p. 38.

[5] Harkness J, Wagner J. The biology and medicine of rabbits and rodents. 4th edition. Philadelphia: Williams & Wilkins; 1995.

[6] Green H, Strauss J. Multiple primary tumors in the rabbit. Cancer 1949;2:673–91.

[7] DiGiacomo RF, Mare CJ. Viral diseases. In: Manning P, Ringler D, Newcomber C, editors. The biology of the laboratory rabbit. 2nd edition. New York: Academic Press; 1994. p. 171–204.

[8] Jenkins JR. Skin disorders of the domestic rabbit. Vet Clin Exot Anim 2001;4:543–64.

[9] Percy DH, Barthold SW. Pathology of laboratory rodents and rabbits. 2nd edition. Ames (IA): State University Press; 2001.

[10] LaRegina MC, Wightman SR. Thyroid papillary adenoma in a guinea pig with signs of cervical lymphadenitis. J Am Vet Med Assoc 1979;175:969–71.

[11] Zarrin K. Thyroid adenocarcinoma of a guinea pig: a case report. Lab Anim 1974;8:145–8.

[12] Olson LD. Experimental induction of cervical lymphadenitis in guinea pigs with group C streptococci. Lab Anim 1976;10:223–31.

[13] Ganaway JR, Allen AM. Obesity predisposes to pregnancy toxemia (ketosis) of guinea pigs. Lab Anim Sci 1971;21:40–4.

[14] Keller LSF, Lang CM. Reproductive failure associated with cystic rete ovarii in the guinea pigs. Vet Pathol 1987;24:335–9.

[15] Christensen HE, Wanstrup J, Ranlov P. The cytology of the Foa-Kurloff reticular cells of the guinea pig. Acta Pathol Microbiol Scand Suppl 1970;212:15–24.

[16] Russo J, Russo IH. Atlas and histologic classification of tumors of the rat mammary gland. J Mammary Gland Biol Neoplasia 2000;5:187–200.

[17] Pliss GB. Pathology of tumours in laboratory animals. Tumours of the rat. Tumours of the auditory sebaceous glands. IARC Sci Publ 1990;99:37–46.

[18] Abbott DP, Prentis DE, Cherry CP. Mononuclear cell leukemia in aged Sprague-Dawley rats. Vet Pathol 1983;20:434–9.

[19] Wagner JE, Owens DR, LaRegina MC, et al. Self trauma and *Staphylococcus aureus* in ulcerative dermatitis of rats. J Am Vet Med Assoc 1977;171:839–41.

[20] Ellis C, Mori M. Skin diseases of rodents and small exotic mammals. Vet Clin Exot Anim 2001;4:493–542.

[21] Greenacre CB. Spontaneous tumors of small mammals. Vet Clin Exot Anim 2004;7:627–51.

[22] Barthold SW, Bhatt PN, Johnson EA. Further evidence for papovavirus as the probable etiology of transmissible lymphoma of Syrian hamsters. Lab Anim Sci 1987;37:283–8.

[23] Garner MM, Raymond JT, Toshkov I, et al. Hepatocellular carcinoma in black-tailed prairie dogs (*Cynomys ludivicianus*): tumor morphology and immunohistochemistry for hepadenavirus core and surface antigens. Vet Pathol 2004;41:353–61.

[24] Phalen DN, Antinoff N, Fricke ME. Obstructive respiratory disease in prairie dogs with odontomas. Vet Clin Exot Anim 2000;3:513–7.

[25] Lightfoot TL. Therapeutics of African pygmy hedgehogs and prairie dogs. Vet Clin Exot Anim 2000;3:155–72.

# Cytologic Diagnosis of Diseases of Hedgehogs

Carles Juan-Sallés, DVM, DACVP[a],*,
Michael M. Garner, DVM, DACVP[b]

[a]ConZOOlting Wildlife Management, La Canyella 1,
Nau 1, 08445 Samalús (Barcelona), Spain
[b]Northwest ZooPath, 654 West Main, Monroe, WA 98296, USA

This article discusses cytologic diagnosis of diseases of hedgehogs. Included are neoplasia, bacterial infections, hepatic lipidosis, extramedullary hematopoiesis, and cardiomyopathy.

**Neoplasia**

Neoplastic diseases are common in hedgehogs according to the literature and the authors' experience. Cytology can prove to be a highly useful diagnostic tool in hedgehog and tenrec medicine. Most reports available in the literature describe neoplasia in African hedgehogs (*Atelerix albiventris*), which are well known for their high incidence of neoplasms in captivity [1–7]. Prevalence of neoplasia in African hedgehogs may exceed 50% in some case series, and the occurrence of different neoplasms in a single animal is not uncommon [6,8]. Malignant neoplasms are particularly common in the literature and the authors' practices for African hedgehogs; 85% of the tumors were classified as malignant in a large case series [6]. A similar trend for neoplastic diseases may occur in other hedgehog species and tenrecs according to the authors' practices, but much smaller numbers of case submissions for these species were available at the time of this writing.

The most common neoplasms in African hedgehogs are oral squamous cell carcinoma, mammary adenocarcinoma, and hematopoietic neoplasia [5,6]. Cutaneous mast cell tumors, soft tissue sarcomas, and endocrine system neoplasms, however, such as adrenocortical carcinoma and thyroid

---

* Corresponding author.
  *E-mail address:* patologia@conzoolting.com (C. Juan-Sallés).

carcinoma, are also seen somewhat frequently [6,8]. All of these tumors may be amenable to cytologic diagnosis.

## Oral squamous cell carcinoma

Squamous cell carcinoma arising from the oral mucosa is one of the most frequent diseases in captive African hedgehogs at the authors' practices. Affected hedgehogs present typically with an oral mass that invades the oral soft tissues and bone, and may become visible externally as a mandibular-maxillary swelling. Squamous cell carcinoma in hedgehogs is slow to metastasize, with pulmonary metastases present in a few cases. Grossly and radiographically, it may resemble bacterial mandibular osteomyelitis as previously described in this species [9]. Cytologic examination of this neoplasm typically reveals numerous neoplastic squamous cells, often polyhedral to elongated (spindeloid), with a moderate to abundant amount of cytoplasm and perinuclear small clear vacuoles or pallor. Anisocytosis, anisokaryosis, and pleomorphism are usually moderate or high, and mitoses may be frequent. Because oral squamous cell carcinomas are commonly ulcerated, suppurative inflammation may be prominent cytologically (Fig. 1). Importantly in this regard, chronic hyperplastic gingivitis associated with bacterial infection is frequent in African hedgehogs and squamous cell hyperplasia and dysplasia in these cases may make it difficult to differentiate squamous cell carcinoma from chronic hyperplastic gingivitis cytologically.

## Mammary gland adenocarcinoma

Mammary neoplasms are very frequent in African hedgehogs and typically correspond to mammary gland adenocarcinoma or solid carcinomas [5,6]. A single or multiple firm subcutaneous mass in the cranial or caudal ventral abdomen may be detected on clinical examination. Malignant mammary gland

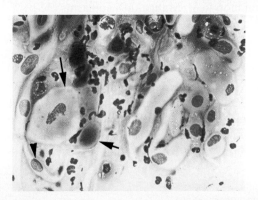

Fig. 1. African hedgehog, squamous cell carcinoma, oral cavity. Note mixture of squamous cells (*long arrow*), atypical parabasal cell (*short arrow*), numerous other parabasal cells, few fibroblasts (*arrowhead*), mixed inflammation, and cellular debris (Modified Wright-Giemsa).

neoplasms are locally invasive but slow to metastasize in hedgehogs in the authors' practices. The cytologic appearance is as seen with most other mammary gland adenocarcinomas in other species, with highly variable numbers of clustered, neoplastic epithelial cells that can be observed on cytologic examination, frequently associated with inflammation or proteinaceous secretions (see Fig. 2 in the article on rabbits elsewhere in this issue).

*Lymphosarcoma*

Lymphosarcoma is the most common hematopoietic neoplasm of African hedgehogs [6]. Gastrointestinal or multicentric lymphosarcomas with involvement of lymph nodes, viscera, and bone marrow are common presentations [4,6]. Cytologically, a uniform and dense population of round cells with scant to moderate amounts of dark blue cytoplasm can be observed. Cells frequently have multiple nucleoli or mitotic figures. Extracellular blue bodies are sometimes seen (Fig. 2). Anaplasia of neoplastic lymphocytes may be low in well-differentiated, small cell lymphosarcoma and diagnosis may be challenging by cytologic examination alone in these cases, although this form of lymphoid malignancy is uncommon in hedgehogs.

*Mast cell tumor*

Mast cell tumors involving the skin of hedgehogs and tenrecs are occasionally seen in the authors' practices. They can be solitary or multiple and often arise from the skin in the head, neck, and axillary region. Mast cell tumors can metastasize to regional lymph nodes in these species. Purple granules in the cytoplasm of neoplastic round cells and in the background may be observed cytologically and usually are accompanied by large numbers of eosinophils (Fig. 3).

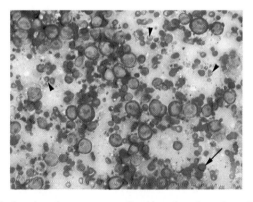

Fig. 2. African hedgehog, lymphosarcoma, popliteal lymph node aspirate. Note large numbers of lymphoid cells with blast-like features, including increased cytoplasmic basophilia, multiple large nucleoli, and some extracellular blue bodies (*arrowheads*). Few small lymphocytes are also present (*arrow*) (Modified Wright-Giemsa).

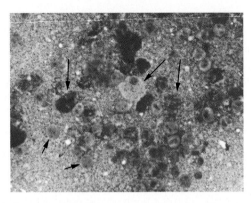

Fig. 3. African hedgehog, mast cell tumor, skin of neck region. Note the variably granulated mast cells (*long arrows*), numerous eosinophils (*short arrows*), and granular matrix indicative of mast cell degranulation (Modified Wright-Giemsa).

## Soft tissue spindle cell sarcomas

A variety of spindle cell sarcomas are also frequent neoplasms in hedgehogs in the literature and authors' practices, especially fibrosarcomas (oral and cutaneous or subcutaneous); uterine leiomyosarcomas; and high-grade, undifferentiated sarcomas [1,2,6,10]. Aspirates from these neoplasms have highly variable cellularity, and also have variable degrees of cellular anaplasia, depending on neoplastic grade (Figs. 4–6). Typically, these tumors are invasive but slow to metastasize.

## Adrenocortical carcinomas

Adrenocortical carcinoma is one of the more frequent neoplasms of the endocrine system in African hedgehogs in the authors' practices. They typically present as large abdominal masses involving one of the adrenal glands

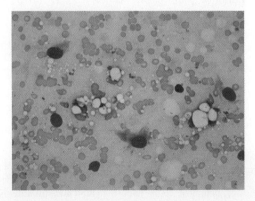

Fig. 4. African hedgehog, low-grade fibrosarcoma, skin of face. Note poorly exfoliative preparation, with few spindle cells containing bland nuclei with single nucleoli and basophilic cytoplasm (Modified Wright-Giemsa).

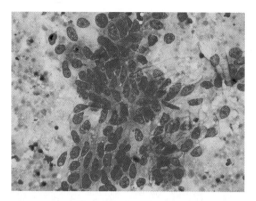

Fig. 5. African hedgehog, intermediate-grade fibrosarcoma, skin of thorax. Note moderately exfoliative preparation with moderate variation in cellular and nuclear size and shape, and cytoplasmic basophilia (Modified Wright-Giemsa).

and apparently are not associated with clinical or morphologic evidence suggesting they are functional. Metastatic disease is common and involves the visceral and parietal peritoneum (seeding metastases) and viscera (liver, spleen, lung). Hemorrhagic effusions may be observed in the abdominal cavity. Cytologic examination of such effusions and aspirates obtained from the masses reveals moderate or numerous round-to-polyhedral cells with a moderate amount of pale basophilic cytoplasm occasionally with a few clear cytoplasmic vacuoles, respectively (Fig. 7).

*Thyroid neoplasia*

Thyroid carcinomas and c-cell carcinomas can be observed in hedgehogs [6,11] and tenrecs, and have cell morphology typical of thyroid neoplasms in other mammals.

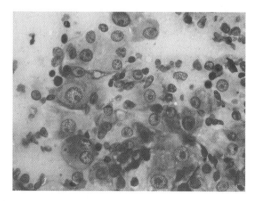

Fig. 6. African hedgehog, high-grade undifferentiated sarcoma, skin of neck region. Note considerable anaplasia characterized by high degree of variability in cellular and nuclear size and shape, large nucleoli, and few binucleated cells.

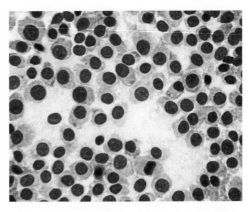

Fig. 7. African hedgehog, adrenocortical carcinoma. Note high cellularity of round-to-polyhedral cells with a moderate amount of cytoplasm; mild anisocytosis; anisokaryosis and anaplasia; and rare, small clear cytoplasmic vacuoles (Modified Wright-Giemsa).

*Other neoplasms*

Uterine and ovarian carcinomas and carcinosarcomas, ovarian granulosa cell tumors, leukemia, and hemangiosarcomas are among other less common neoplasms in African hedgehogs in the authors' practices.

**Bacterial infections**

Bacterial infections seem not to be common in hedgehogs and tenrecs in the authors' practices, except for bacterial gingivitis associated with periodontal disease, which in some cases may progress to mandibular-maxillary osteomyelitis and abscesses [9]. Animals may present with oral, facial, or maxillary-mandibular swelling or masses that can resemble oral squamous cell carcinoma grossly; cytologic examination and, if necessary, biopsy are required to distinguish these diseases. Cytologically, aspirates may be consistent with an abscess or with pyogranulomatous inflammation characterized by the presence of abundant neutrophils and fewer macrophages and multinucleate giant cells. Bacteria may be observed and a Gram stain may be helpful in characterizing intralesional bacteria before aerobic and anaerobic culture results are available. With pyogranulomatous inflammation, neutrophils may clump around bacterial colonies, as occurs with *Actinomyces* infection (Fig. 8) [9]. In early stages of periodontal disease associated with gingivitis and bacterial colonization, gingival hyperplastic lesions with dysplasia of epithelial cells may be prominent and resemble cytologically an oral squamous cell carcinoma and caution must be taken when evaluating a cytologic preparation from such lesions.

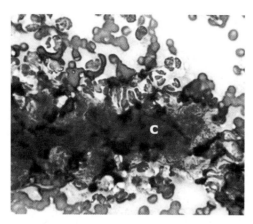

Fig. 8. African hedgehog, associated jaw lesion. Note clusters of viable and degenerative neutrophils oriented around a dense cluster of filamentous bacteria (*c*) (Modified Wright-Giemsa).

Other bacterial infections that may be diagnosed cytologically in hedgehogs are suppurative or pyogranulomatous osteomyelitis in the appendicular skeleton, intra-abdominal abscesses, and suppurative or pyogranulomatous pneumonia.

## Hepatic lipidosis

According to the literature [7] and authors' practices, hepatic lipidosis is a frequent disease in captive hedgehogs and is commonly associated either with cardiomyopathy or malignant neoplasms in these species but may also be seen with endocrine or metabolic derangements, anorexia, obesity, and pregnancy stressors. Fine-needle aspirates from enlarged livers reveal hepatocytes with cytoplasmic variably sized clear vacuoles, with displacement and compression of the nucleus to the periphery in some cases. Occasional hepatocytes have cytoplasmic dark (hemosiderin) or green (bile) pigment (Fig. 9).

## Extramedullary hematopoiesis

As with other small mammals, extramedullary hematopoiesis is frequent in captive hedgehogs and its etiology is undetermined, although it may be reactive to anemia and infection in some cases [7]. This lesion typically involves the spleen and causes splenomegaly. Fine-needle aspirates from enlarged spleens reveal erythroid, myeloid, and megakaryocytic precursors, and few resident lymphocytes, plasma cells, and macrophages (Fig. 10).

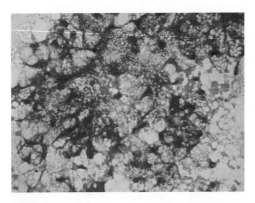

Fig. 9. African hedgehog, hepatic lipidosis. Note cytoplasm of all hepatocytes is distended with variably sized spherical clear vacuoles consistent with lipid, and nuclei are eccentrically displaced (Modified Wright-Giemsa).

## Cardiomyopathy

Cardiomyopathy is one of the major diseases of captive African hedgehogs [12]. Animals may die unexpectedly with no premonitory signs or develop congestive heart failure. A few hedgehogs present with hydrothorax or ascites; aspirates of body cavity effusions in these cases typically reveal a modified transudate, but underlying diseases other than cardiomyopathy must be ruled out. Direct smears generally have low cellularity and may be bloody. Cytocentrifuged preparations are best for establishing different cell populations. Typically, these preparations have a mixture of well-preserved macrophages, neutrophils, lymphocytes, plasma cells, and erythrocytes. The macrophages sometimes contain phagocytized cellular debris or

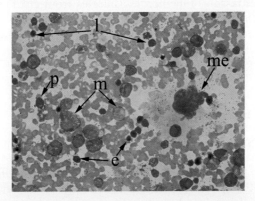

Fig. 10. African hedgehog, splenic extramedullary hematopoiesis. Note numerous erythrocytes, myeloid precursors (*m*), erythroid precursors (*e*), megakaryocyte (*me*), plasma cells (*p*), and lymphocytes (*l*) (Modified Wright-Giemsa).

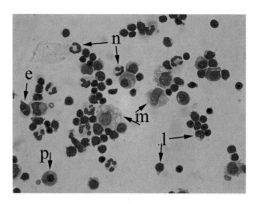

Fig. 11. African hedgehog, modified transudate, thoracic cavity. Note viable neutrophils with no toxic change (n), macrophages (m), plasma cells (p), eosinophils (e) and few erythrocytes. One macrophage contains a phagocytized erythrocyte (Cytospin preparation, modified Wright-Giemsa).

dark-staining pigment (usually hemosiderin) (Fig. 11). Hepatic lipidosis is also frequent in these hedgehogs and can be diagnosed cytologically, as described previously.

## References

[1] Greenacre CB. Spontaneous tumors of small mammals. Vet Clin Exot Anim 2004;7:627–51.
[2] Ramos-Vara JA. Soft tissue sarcomas in the African hedgehog (*Atelerix albiventris*): microscopic and immunohistologic study of three cases. J Vet Diagn Invest 2001;13:442–5.
[3] Ramos-Vara JA, Miller MA, Craft D. Intestinal plasmacytoma in an African hedgehog. J Wildl Dis 1998;34:377–80.
[4] Raymond JT, Clarke KA, Schafer KA. Intestinal lymphosarcoma in captive African hedgehogs. J Wildl Dis 1998;34:801–6.
[5] Raymond JT, Garner MM. Mammary gland tumors in captive African hedgehogs. J Wildl Dis 2000;36:405–8.
[6] Raymond JT, Garner MM. Spontaneous tumours in captive African hedgehogs (*Atelerix albiventris*): a retrospective study. J Comp Pathol 2001;124:128–33.
[7] Raymond JT, White MR. Necropsy and histopathologic findings in 14 African hedgehogs (*Atelerix albiventris*): a retrospective study. J Zoo Wildl Med 1999;30:273–7.
[8] Juan-Sallés C, Raymond JT, Sarria-Perea JA, et al. Neoplasia in captive African hedgehogs (*Atelerix albiventris*) with emphasis on two adrenocortical carcinomas. Vet Pathol 2001; 38:585.
[9] Martínez LS, Juan-Sallés C, Cucchi-Stefanoni K, et al. *Actinomyces naeslundii* infection in an African hedgehog (*Atelerix albiventris*) with mandibular osteomyelitis and cellulitis. Vet Rec 2005;157:450–1.
[10] Peauroi JR, Lowenstine LJ, Munn RJ, et al. Multicentric skeletal sarcomas associated with probable retrovirus particles in two African hedgehogs (*Atelerix albiventris*). Vet Pathol 1994;31:481–4.
[11] Miller DL, Styer EL, Stobaeus JK, et al. Thyroid c–cell carcinoma in an African pygmy hedgehog (*Atelerix albiventris*). J Zoo Wildl Med 2002;33:392–6.
[12] Raymond JT, Garner MM. Cardiomyopathy in captive African hedgehogs (*Atelerix albiventris*). J Vet Diagn Invest 2000;12:468–72.

# Cytologic Diagnosis of Diseases of Ferrets

Pauline M. Rakich, DVM, PhD[a,b],*,
Kenneth S. Latimer, DVM, PhD[a]

[a]*Department of Pathology, College of Veterinary Medicine, University of Georgia, Athens, GA 30602, USA*
[b]*Athens Veterinary Diagnostic Laboratory, College of Veterinary Medicine, University of Georgia, Athens, GA 30602-7383, USA*

Ferrets have become an extremely popular pet in the United States as the population has become increasingly urbanized. The American Ferret Association estimates that there are currently some 7 to 10 million ferrets in the United States [1]. They are curious, playful, easily cared for animals that can be kept as pets by people living in apartments that do not allow dogs and cats. In addition, they provide an alternative indoor pet for those who are allergic to cats. The small body size and long narrow shape make them the near-perfect subject for cytology because nearly all tissues and organs are accessible by fine-needle aspiration. Their active, somewhat fractious natures may increase the risk of adverse events, such as hemorrhage and trauma, during sampling; but with adequate manual or chemical restraint, any complications should be prevented.

Specific information concerning cytology of ferrets is limited in the literature, but much of the well-established cytologic features of lesions in dogs and cats are directly applicable to diseases of ferrets. The first objective of cytologic evaluation of tissues and lesions is to determine whether a lesion is inflammatory or noninflammatory. Noninflammatory conditions can further be classified as neoplastic and nonneoplastic lesions. Mixtures of inflammatory cells and tissue cells present a diagnostic challenge because it may be difficult to differentiate inflammation with reactive tissue proliferation from a neoplasm with inflammation. In some cases, biopsy and histopathology are necessary to make this distinction accurately.

---

* Corresponding author.
*E-mail address:* rakich@vet.uga.edu (P.M. Rakich).

## Inflammation

Inflammatory lesions are characterized by the predominant cells comprising the infiltrate. These most commonly include purulent, pyogranulomatous, granulomatous, and lymphocytic inflammation. The type of inflammation is significant because it frequently provides a clue to the cause of the inflammation. Purulent exudates usually occur with infectious lesions, particularly bacterial infections. Consequently, smears composed of numerous neutrophils should always be searched carefully for bacteria. Depending on the particular bacterium involved, neutrophils may be nondegenerate or exhibit varying degrees of degeneration. They are recognized as degenerate when their nuclei are swollen, smooth, and pink as compared with tight, dense, and blue-purple nuclei of normal neutrophils. Degenerate neutrophils indicate a toxic environment produced by the bacteria.

Macrophages are usually numerous or predominate in lesions caused by fungi, foreign material, protozoa, and mycobacteria. Monocytes that are recruited from the blood and reach an inflammatory site undergo activation and become macrophages. Macrophages are larger than neutrophils and have an oval or reniform nucleus; fine chromatin; small or inconspicuous nucleoli; and moderate to abundant pale cytoplasm, which is frequently vacuolated. They commonly contain phagocytosed cell debris and sometimes infectious agents, because their function is to remove cell debris in tissues with inflammation or necrosis. Fusion of macrophages produces multinucleate giant cells, which have multiple nuclei and more abundant cytoplasm. These cells also commonly contain phagocytosed material within their cytoplasm.

A predominance of eosinophils is usually associated with hypersensitivity reactions; parasitism; and rare infectious agents (eg, *Pythium* sp). Keratin and hair fragments from ruptured hair follicles in the dermis commonly produce an intense foreign body inflammatory reaction that contains many eosinophils.

## Infection

A variety of agents can be seen in infectious lesions and they can be identified tentatively or definitively by cytologic examination. Bacteria are the most common organisms observed in cytologic specimens. With Romanowsky's-type stains, including Diff-Quik, most bacteria appear dark blue, as compared with blue-purple nuclei and nuclear debris. Bacteria may be cocci, rods, coccobacilli, and filaments. Bacteria can be seen within neutrophils or in the background of the smear. The notable exceptions are mycobacteria, which fail to stain with Romanowsky's stains because of the high fat content of their cell walls. They appear in smears as clear, rod-shaped structures within macrophages or multinucleate giant cells or free in the background of the smear among the inflammatory cells (Fig. 1).

Fungi occur as hyphae or yeasts and either form can be seen in smears from fungal infections, depending on the particular organism involved. Fungal hyphae are linear structures, usually measuring approximately 5 to 10 µm in width, that may have internal septations and branching. Most fungi stain basophilic, but some fungi stain poorly or not at all and are only visible as unstained linear "ghosts" among inflammatory cells (Fig. 2). Some fungi are pigmented (dematiaceous) and appear various shades of gold, brown, or black. A few fungi and fungus-like organisms are commonly associated with an eosinophilic inflammatory response. Yeasts are round to oval structures that stain basophilic typically and sometimes have a clear cell wall or thick clear capsule. They vary greatly in size, from several micrometers (eg, *Candida*) to greater than 200 µm (eg, *Coccidioides immitis*).

## Neoplasia

Although nonneoplastic and inflammatory lesions occur in ferrets, the incidence of neoplastic lesions seems to be much higher. One author has stated that most ferrets develop one or more neoplasms by the age of 5 years [2]. When evaluating neoplasms, the first step is to classify them as epithelial or mesenchymal/connective tissue. Mesenchymal tumors are frequently further subdivided into round cell tumors and spindle cell tumors. In the case of many well-differentiated tumors, a definitive diagnosis may be made by cytologic examination. Less well-differentiated tumors may be identified as epithelial or mesenchymal tumors, but a more definitive diagnosis frequently requires histologic examination. Some tumors may be so poorly differentiated that they can be recognized as malignant because of obvious

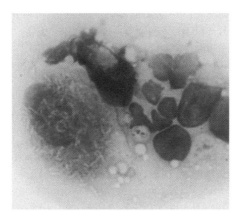

Fig. 1. Macrophage with numerous intracytoplasmic clear, unstained, rod-shaped structures typical of mycobacteria.

Fig. 2. Unstained "ghost" of a fungal hypha surrounded by inflammatory cells.

cytologic features of malignancy, but no further identification is possible. In such instances, biopsy and histopathology are required for diagnosis. Additional diagnostic techniques, such as immunohistochemical staining or electron microscopic examination, also may be required to make a definitive diagnosis histologically.

Epithelial tumors usually exfoliate readily so smears tend to be relatively cellular. The cells are usually distributed singly and in clusters and sheets. The individual tumor cells are usually polyhedral and have round or oval nuclei and moderately abundant cytoplasm. Spindle cell tumors are derived from mesenchymal tissue. These neoplasms are composed of cells that typically are spindle-shaped, stellate, or polyhedral. Nuclei are ovoid to elongate and the cytoplasm is usually pale and merges gradually with the background of the smear so that cell margins are frequently indistinct. Smears from connective tissue tumors usually contain few cells because the supporting stroma surrounding the cells prevents their detachment or exfoliation from the tissue. Round cell tumors are also derived from mesenchymal tissues, but the cells are distinctly round to oval and they generally exfoliate easily and are usually numerous in smears. Diagnosis of round cell tumors is based solely on the appearance of the tumor cells, with tissue architecture being unnecessary for diagnosis. Consequently, the diagnosis of round cell tumors is frequently possible by cytologic examination alone.

In general, benign tumors are recognized by their similarity to the normal tissue from which they are derived. Nuclei are usually small, chromatin is dense, nucleoli are small or unapparent, and cytoplasm is present in amounts typical of normal tissue. Mitotic figures are usually not numerous in benign tumors. Classification of tumors as malignant is based on the presence of morphologic criteria of malignancy, which primarily involve nuclear features. These include monomorphic cell population; cell enlargement; pleomorphism of the cells (variation in size and appearance); anisocytosis (variation in cell size); anisokaryosis (variation in nuclear size); high nuclear/cytoplasmic ratio caused by nuclear enlargement; numerous and abnormal mitotic figures; multinucleation; and variation in the number, size, and shape of nucleoli. Sometimes it is not possible to differentiate a well-differentiated malignancy from a benign tumor or reactive tissue, however, particularly in lesions involving connective tissue. In such cases, biopsy and histopathology are recommended for diagnosis.

**Skin masses**

Tumors of the skin are the third most common neoplasms in ferrets after endocrine and hemolymphatic neoplasms [3]. Consequently, a large number of lesions are easily accessible for sampling and cytologic examination. In general, preparations needed for obtaining samples of skin lesions include manual restraint and surface cleansing with or without shaving. Fine-needle aspiration usually provides more diagnostic specimens than do tissue imprints or scrapings.

Mast cell tumors are a common, if not the most common, skin tumor in ferrets [3–5]. They can be single or multiple and may be present for many months without changing in appearance. Unlike in dogs, mast cell tumors in ferrets behave in a benign manner. Because of biologically active substances in the mast cell granules, however, such as histamine, they can be pruritic causing the animal to scratch and chew at the lesion, which can result in alopecia and ulceration. In contrast to dogs with mast cell tumors, ferrets do not develop gastric ulceration. Clinically, they can be flat, raised, or nodular well-circumscribed firm lesions with a normal skin surface or alopecia and ulceration caused by self-trauma. They can occur anywhere on the body and are most common in ferrets between 2 and 8 years old. Smears from aspirates of mast cell tumors are composed of discrete polyhedral (round) cells with round or oval nuclei; granular chromatin; single small nucleoli; and moderate amounts of cytoplasm with fine purple granules (Fig. 3). Some stains fail to fix the cells quickly enough, however, and

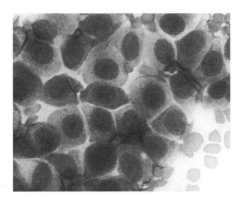

Fig. 3. Mast cells with faint fine purple cytoplasmic granules.

they degranulate before the granules can be stained. In such cases, the cells have a moderate amount of pale blue cytoplasm without any granules or fine clear vacuoles. Although mast cell tumors in dogs typically have numerous eosinophils mixed with the mast cells, those of ferrets and cats usually do not.

Sebaceous gland tumors are also common in ferrets [2,3,6,7]. Both benign and malignant tumors have been reported, although most are benign and consist of sebaceous adenomas and sebaceous epitheliomas. Sebaceous gland tumors can occur anywhere on the body but they may be most common on the head and neck. Grossly, sebaceous gland tumors are firm, raised, warty, or multilobulated pale nodules, usually measuring less than 3 cm in diameter. They may be ulcerated because of scratching and chewing and appear angry and suggestive of a malignancy when, instead, they are typically benign. Smears prepared from aspiration of sebaceous adenomas usually consist of clusters of cells with small dense ovoid nuclei; single nucleoli, which may not be visible because of the dense chromatin; and moderately abundant finely vacuolated cytoplasm. These cells are the mature sebocytes comprising sebaceous glands (Fig. 4). Sebaceous epitheliomas, which occur more commonly in ferrets than sebaceous adenomas, are composed of a combination of mature sebocytes and reserve cells. Reserve cells are the generative population of sebaceous glands and constitute the predominant cell type in sebaceous epitheliomas. Consequently, smears from sebaceous adenomas are composed of two cell types. Most of the cells consist of clusters and sheets of small uniform cells with ovoid nuclei, coarse chromatin, single small or inconspicuous nucleoli, and a small amount of pale blue cytoplasm without evident cell margins. These are the reserve cells of the sebaceous glands. Scattered among these small cells are occasional individual or small clusters of cells with similar nuclei but more abundant vacuolated cytoplasm consistent with mature sebocytes (Fig. 5).

Basal cell tumors are composed of the proliferating pluripotential basal cells of the epidermis. They are common in dogs and cats and they also occur in ferrets, but their incidence in ferrets is difficult to determine because

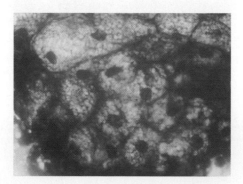

Fig. 4. Mature sebocytes in a sebaceous adenoma.

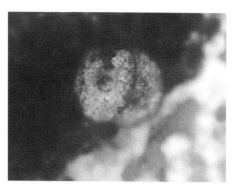

Fig. 5. Cluster of reserve cells and several mature sebocytes from a sebaceous epithelioma.

they have been confused in the old literature with sebaceous epitheliomas [3,6]. They can occur anywhere on the body and are usually firm, pale, plaques, nodules, or pedunculated masses. Like sebaceous epitheliomas, they are benign. They are composed entirely of basal cells, which resemble the reserve cells of sebaceous glands. Smears of basal cell tumors consist of clusters and sheets of small uniform cells with small dense oval nuclei, small nucleoli, and small amounts of basophilic cytoplasm without visible cell margins (Fig. 6). It may not be possible to differentiate a basal cell tumor from a sebaceous epithelioma cytologically, but the distinction is unnecessary clinically because complete surgical excision of both tumors is curative.

Apocrine glands are the sweat glands of the skin, which are associated with hair follicles. Nonneoplastic and benign and malignant tumors of apocrine glands occur in ferrets, although they are not common [3,6,7]. Apocrine cysts are usually firm nodules that are filled with clear colorless or pale yellow fluid. They may collapse completely with aspiration and not recur or slowly enlarge as the gland refills with fluid. Smears prepared from apocrine cysts are typically acellular and the diagnosis is based on

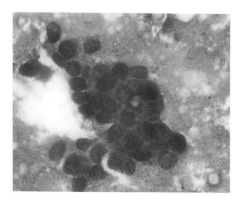

Fig. 6. Cluster of basal cells from a basal cell tumor.

the gross appearance of the fluid; lack of cells; and resolution of the mass with aspiration of the fluid, which resembles water. Aspiration of tumors usually yields more cellular smears. Smears from apocrine adenomas are usually composed of small polyhedral to columnar cells distributed singly and in clusters. They have small oval nuclei, small nucleoli, and moderate amounts of pale blue cytoplasm, which may have blue-gray granules or globules (Fig. 7). Malignant tumors, apocrine adenocarcinomas, tend to produce more highly cellular smears with more prominent variation in nuclear (anisokaryosis) and cell (anisocytosis) size.

Malignant lymphoma or lymphosarcoma (LSA) is a common tumor in ferrets, second only to endocrine tumors, and it is the most common neoplasm of young (<1 year old) ferrets [8]. The skin is not involved, however, as commonly as are lymph nodes and other tissues [9,10]. Skin lesions associated with LSA are variable as in other species and this neoplasm should be kept in mind in any skin lesions that do not respond to antibiotics. Progressive pruritic dermatitis, alopecia, erythema, excoriations, ulcers, crusts, plaques, and nodules have been reported. Lesions are usually inexorably progressive and can involve the entire body, including foot pads and planum nasale. Aspiration of skin lesions is frequently nondiagnostic, particularly if the lesions are inflamed because of ulceration or infection or when no distinct nodules are present. When well-defined skin nodules are present, aspiration can be diagnostic. The lesions are composed of a uniform population of a single type of lymphocytes (ie, monomorphic population), which are usually large immature lymphocytes in young ferrets and more commonly small mature lymphocytes in adult ferrets. The size of lymphocytes can be determined by comparing their nuclei with red blood cells (RBCs). Small lymphocytes have nuclei that measure approximately the size of RBCs with dense chromatin, no visible nucleoli, and a very thin or incomplete rim of cytoplasm. Medium-size lymphocytes have nuclei that measure two times the diameter of RBCs. Their chromatin is less dense, nucleoli are frequently visible, and they have a small amount of dark blue cytoplasm. Large

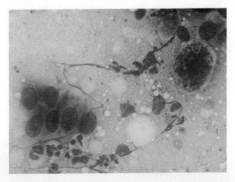

Fig. 7. Cluster of apocrine columnar epithelial cells with blue-gray secretory granules surrounded by inflammatory cells.

lymphocytes have nuclei that measure three times the diameter of RBCs with fine granular chromatin, one or several prominent nucleoli, and a small to moderate amount of dark blue cytoplasm. When smears of lesions consist of a monomorphic population of lymphocytes, either small lymphocytes or medium to large lymphocytes, LSA is more likely than inflammation.

Squamous cell carcinoma does not seem to be as common in ferrets as in dogs and cats. The tumors may be single or multiple and occur anywhere in the skin but perhaps more commonly on the jaw. The tumors are pale, firm nodules, plaques, or ulcerated lesions. They are typically locally invasive and may metastasize. The prognosis is considered poor, particularly when there is invasion into underlying bone. Aspiration is recommended over impression smears because ulcerated lesions are commonly secondarily infected and inflamed, which interferes with interpretation because inflammation and ulceration of the skin result in proliferation of squamous cells, which can be difficult to differentiate from neoplasia. Smears from well-differentiated squamous cell carcinomas consist of large polyhedral cells distributed singly and in clusters and sheets (Fig. 8). The cells have round or oval nuclei that vary moderately in size, visible nucleoli that may be multiple or vary in size and shape, and moderate to abundant pale blue cytoplasm. Binucleate and multinucleate cells are frequently common. Some cells may be more irregularly shaped with prominent tails. With keratinization, nuclei become shrunken and dense (pyknotic) or completely disappear and cytoplasm becomes darker with prominent angular margins. Inflammatory cells are frequently also present and include lymphocytes, plasma cells, and neutrophils.

A variety of spindle cell tumors occur in ferrets but they are not as common as those discussed previously. These include fibroma, fibrosarcoma, myxoma, myxosarcoma, leiomyoma, leiomyosarcoma, chordoma, synovial cell sarcoma, hemangioma, and hemangiosarcoma [3]. Unlike many epithelial tumors and round cell tumors, a specific diagnosis is rarely made cytologically. Furthermore, in many instances it is not possible to distinguish connective tissue hyperplasia from connective tissue neoplasm by

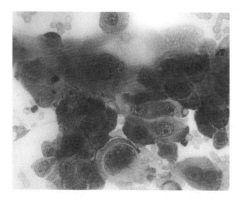

Fig. 8. Squamous cells from a squamous cell carcinoma with several multinucleate tumor cells.

cytology. Diagnosis frequently requires biopsy and histologic examination to evaluate architecture of the tissue to distinguish reactive from neoplastic tissue. These tumors are composed of cells that can be spindloid, stellate, or polyhedral in some cases. Nuclei are usually oval to elongate and cytoplasm appears wispy and merges into the background of the smear. These tumors commonly have relatively abundant supporting stroma that interferes with cell detachment (exfoliation), usually resulting in sparsely cellular smears. This is particularly true of very firm or hard tumors, which contain large amounts of collagen, cartilage, or bone, such as chordomas, which are a relatively common tumor on the tail of ferrets.

Inflammatory skin lesions are not as common in ferrets as neoplastic ones. Bacterial skin infections usually result from bite wounds occurring during play, fighting, and mating or from puncture wounds [4]. A variety of gram-positive and gram-negative organisms have been cultured from ferrets with dermatitis, abscesses, and cellulitis. Smears can be made from aspiration of lesions or from draining exudate. Smears prepared from bacterial lesions typically consist of neutrophils, cell debris, macrophages, lymphocytes, and plasma cells. The neutrophils may appear nondegenerate or exhibit varying degrees of degeneration, which reflects a toxic environment produced by the bacteria. Degeneration is characterized by swollen, smooth pink chromatin, which is in contrast to the normal dense dark blue-purple compact chromatin of normal neutrophils in smears. The presence of neutrophils in a smear should prompt a search for infectious agents, which may be present in very small numbers or in large numbers, both within neutrophils and free in the background of the smear. Depending on the duration of the lesion, connective tissue spindle cells may also be present.

Lumpy jaw, or actinomycosis, is an infection caused by *Actinomyces* sp, which is a gram-positive filamentous bacterium. This infection is characterized by development of firm nodules, frequently with draining tracts, which may contain yellow grains called "sulfur granules" in the cervical region [11]. Infection with this agent is thought to result from damage to the oral mucosa because it is commonly a normal component of the oral flora. This infection has been reported rarely in ferrets but a tentative diagnosis is possible with cytologic examination. The nodules must be differentiated from neoplasms and other infectious conditions. The demonstration of large masses of bacteria that are beaded filaments is typical of *Actinomyces* sp and *Nocardia* sp and specific identification requires culture. Because the two organisms respond to different antibiotics, culture is also recommended for appropriate antibacterial therapy.

Ferrets are susceptible to various strains of human, avian, bovine, and opportunistic mycobacteria and cytology is an effective means of making a preliminary diagnosis of mycobacteriosis. Clinical signs include weight loss, anorexia, hepatosplenomegaly, intestinal nodules, abdominal lymphadenopathy, and localized lesions [11]. Two adult ferrets with infection with *Mycobacterium genavense* presented with generalized peripheral lymphadenopathy and conjunctival swelling caused by granulomatous inflammation [12]. Smears prepared from aspiration of lesions caused by mycobacterial infection usually contain a predominance of large macrophages that resemble epithelial cells (epithelioid macrophages) and variable numbers of multinucleate giant cells and neutrophils. Depending on the particular strain of mycobacteria, the organisms may be rare or numerous and are seen as unstained clear rod-shaped structures within macrophages (see Fig. 1). In infections with numerous organisms, bacteria are also present in the background of the smear.

Fungal infections of the skin of ferrets are uncommon. Dermatophytosis, or "ringworm," is usually caused by *Microsporum canis* or *Trichophyton mentagrophytes* and is associated with overcrowding or exposure to cats, the natural host of *M canis* [13]. Infection usually occurs in young animals. The lesions are similar to those in dogs and cats with dermatophytosis (ie, circular expanding foci of alopecia, brittle or broken hairs, and hyperkeratosis or crusting). Smears from crusty lesions usually consist of cell debris, keratin, neutrophils, macrophages, lymphoid cells, and sometimes arthrospores (Fig. 9). The diagnosis is not usually confirmed cytologically, however, and culture is recommended.

Infections with other fungi have been reported rarely in ferrets [4,13], but fungal infection should be considered for nodular draining lesions that do not respond to antibiotic treatment. Blastomycosis, coccidioidomycosis, cryptococcosis, and mucormycosis have been reported in ferrets [4,13]. Aspirates of lesions caused by fungal organisms typically contain macrophages, neutrophils, and multinucleate giant cells with fewer lymphocytes and plasma cells. Organisms may be numerous or rare and sometimes extensive search is necessary to find agents. *Blastomyces dermatitidis* is a round basophilic yeast that measures approximately 6 to 20 µm in diameter (about the size of a RBC to slightly larger than a neutrophil) with a thick clear

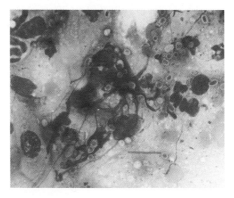

Fig. 9. Exudate containing arthrospores from a case of dermatophytosis caused by *Microsporum canis*.

refractile cell wall (Fig. 10). The organisms may be phagocytosed within macrophages and multinucleate giant cells or scattered free among the cells.

Salivary gland mucocele is a nonneoplastic, noninflammatory lesion that has been reported rarely in ferrets [14,15]. The lesions develop as a swelling on the side of the face, orbital area, and commissures of the mouth but other sites are possible depending on the salivary gland involved. The masses tend to be fluctuant and aspiration yields thick, viscous, or mucinous clear, blood-tinged, or pale yellow fluid. Smears prepared from the fluid consist of mucin, which may appear as variably sized masses of pale blue smooth (hyaline) material or pink to blue granular to fibrillar material in the background (Fig. 11). Cellularity is variable. It is generally low initially and primarily composed of vacuolated macrophages. With chronicity, inflammation increases and neutrophils become progressively more numerous. RBCs are also present usually, and with chronicity macrophages phagocytose the RBCs and degrade them to produce hemosiderin and hematoidin. Hemosiderin is a blue-gray pigment consisting of the iron derived from the RBCs and hematoidin is a yellow-gold crystalline pigment that represents tissue bilirubin (Fig. 12).

**Lymph nodes**

The indications for lymph node aspiration are to determine the cause of lymph node enlargement and to evaluate a regional lymph node for the presence of metastatic tumor cells. Ferrets have abundant axillary, inguinal, and popliteal fat, which can be misinterpreted as lymph node enlargement. This perinodal fat also makes lymph node aspiration difficult and smears frequently consist entirely of subcutaneous fat and are consequently

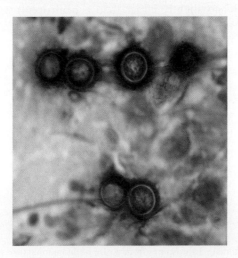

Fig. 10. *Blastomyces dermatitidis* yeasts.

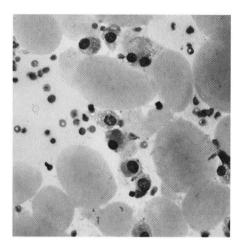

Fig. 11. Masses of basophilic mucin and scattered macrophages from a sialocele.

nondiagnostic. The most common cause of lymphadenopathy in ferrets is LSA, although it does not occur as commonly as in dogs with LSA [8]. The most common form of LSA involving the lymph nodes in ferrets occurs in older animals and it is usually the lymphocytic form, which consists of small, well-differentiated lymphocytes. Aspiration of normal lymph nodes yields a mixture of small, medium, and large lymphocytes and plasma cells. Small lymphocytes predominate in normal lymph nodes. Consequently, diagnosis of the lymphocytic form of LSA is difficult cytologically because the cells appear normal. It is not until the advanced stages, when the cells virtually replace the node, that a diagnosis is possible cytologically because the smears consist entirely of a monomorphic population of small lymphocytes.

Regional lymph nodes may be enlarged when there is a tumor in the tissue drained by that lymph node. The lymphadenopathy may occur because of inflammation secondary to tumor necrosis, ulceration, and infection;

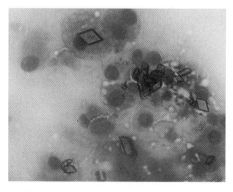

Fig. 12. Numerous hematoidin crystals.

reaction to tumor antigens; or because of metastasis of tumor cells to the lymph node. Aspiration and cytologic examination of the enlarged lymph nodes can help determine the cause of the enlargement. If a regional lymph node is not enlarged, detection of metastasis is unlikely by cytology. One malignancy that has been reported to metastasize commonly in ferrets is squamous cell carcinoma. Smears from a lymph node with metastatic squamous cell carcinoma typically contain neoplastic cells as previously described mixed with variable numbers of mixed lymphocytes and plasma cells and frequently also neutrophils. In advanced cases when the neoplastic cells have replaced normal lymph node tissue, aspirates may consist only of tumor cells.

## Body cavity effusions

Free fluid is not normally present in the thoracic and abdominal cavities of ferrets. Aspiration of fluid from body cavities can be done for both diagnostic and therapeutic purposes, particularly in the thoracic cavity where the presence of even small amounts of fluid can interfere with pulmonary function. Effusions are uncommon in ferrets. The most frequent cause of effusions in ferrets is LSA in young animals with mediastinal LSA [8]. Despite the presence of a thymic mass and pleural fluid, the affected animals frequently do not exhibit dyspnea because of the ability to compensate until late in the course of the disease. Smears prepared from the aspirated pleural fluid are usually moderately to highly cellular and composed of a uniform population of large immature lymphocytes (Fig. 13). Ferrets with heartworm disease also can develop pleural effusion [16,17]. In such instances the fluid may contain numerous eosinophils.

Abdominal effusions occur in adult ferrets with cardiac disease. These fluids usually have a low cell count ($<5000/\mu L$) composed of mixed cells. The cells include macrophages, neutrophils, lymphocytes, and mesothelial

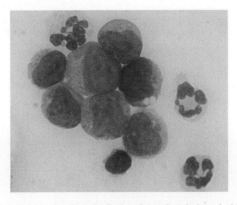

Fig. 13. Large lymphocytes in pleural effusion from thymic lymphoblastic lymphoma.

cells. Macrophages typically contain phagocytosed RBCs or hemosiderin within their cytoplasm. Mesothelial cells have round or oval nuclei; fine chromatin; a single nucleolus; and moderate amounts of medium to dark blue cytoplasm, which may have a pink fringe. The cells are distributed singly and in clusters. Mesothelial cells undergo hypertrophy and hyperplasia with effusions and exhibit variable anisocytosis and anisokaryosis (Fig. 14). Activated mesothelial cells phagocytose material and can be difficult to differentiate from macrophages.

Mesothelioma has been reported in ferrets [2]. This is a rare tumor of the cells lining the abdominal, pericardial, and pleural cavities. Individuals with mesothelioma usually develop an effusion in the affected body cavity. The diagnosis of mesothelioma is difficult because mesothelial cells typically proliferate whenever there is fluid in a body cavity and they readily exfoliate within the effusion. Consequently, reactive mesothelial cells must be differentiated from neoplastic cells. This distinction may not be possible cytologically.

## Respiratory system

Pneumonia is not common in ferrets and it is usually caused by viral infections (distemper virus and influenza virus). Bacterial infections are usually secondary to aspiration or to a primary viral infection [18]. Various bacteria have been reported. Tracheal washes are useful for cytology and culture to determine appropriate antibacterial therapy. If the recovered tracheal wash fluid is cloudy or contains clumps of material, direct smears or crush preparations can be made. If the fluid appears clear, however, cellularity is probably low and smears should be prepared from centrifuged sediment to ensure adequate material is present to examine. Smears prepared from tracheal wash smears consist of neutrophils, cell debris, mucin, macrophages, ciliated columnar epithelial cells, and bacteria. In the case of

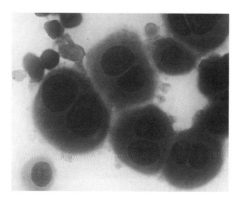

Fig. 14. Reactive mesothelial cells from abdominal effusion.

aspiration pneumonia, foreign material (eg, fragments of plant material) may also be present (Fig. 15).

Lung aspiration is not usually diagnostic unless a distinct pulmonary mass is evident radiographically. Even with diffuse pulmonary disease, tracheal washes are usually more diagnostic than aspirates.

**Spleen**

Splenomegaly is a common finding in ferrets [19]. It occurs in normal ferrets and in ferrets with a variety of conditions, and in most cases the splenomegaly is not clinically significant. Fine-needle aspiration, however, is easily done, usually does not require any sedation, and does not cause any adverse effects. The most common cause of splenic enlargement in ferrets is extramedullary hematopoiesis. Aspirates of smears from spleens with extramedullary hematopoiesis resemble bone marrow aspirates and are composed of mature blood cells, mixed erythroid cells, megakaryocytes, and myeloid cells (Fig. 16). Mixed lymphocytes and plasma cells also may be present because the spleen normally contains lymphoid follicles and plasma cells.

Although the spleen is commonly infiltrated by neoplastic lymphocytes in ferrets with LSA, diagnosis by splenic aspiration may be difficult. In young ferrets, usually <1 year, LSA is usually caused by immature lymphocytes and if the infiltration is extensive enough, the diagnosis is not difficult because the smear contains a predominance of immature lymphocytes instead of the mixed population present in a normal spleen. In adults with the small cell form of LSA, however, the diagnosis is more problematic unless the infiltration is so diffuse that the neoplastic cells have replaced all other cells and the smear consists of a monomorphic population of small lymphocytes.

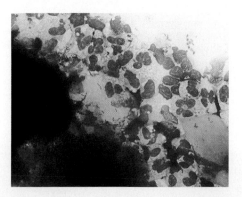

Fig. 15. Smear of tracheal wash with numerous mixed bacteria and large particulate foreign material from an aspiration pneumonia.

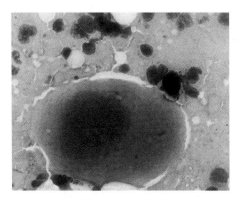

Fig. 16. Megakaryocyte and immature red and white blood cells from splenic extramedullary hematopoiesis.

## Liver

Except for neoplasia, liver disease is not common in ferrets. A variety of neoplasms affect the ferret liver, including primary hepatocellular tumors and metastatic tumors. The most common tumor involving the liver of ferrets is LSA. The diagnosis may not be possible by cytologic examination of aspirates, however, unless the infiltration of neoplastic cells is diffuse enough to be aspirated on a random aspirate and in high enough numbers to be visible among the blood cells that are usually present in liver aspirate smears. Smears prepared from liver aspirates typically consist of abundant blood and variable numbers of hepatocytes distributed singly and in clusters. Hepatocytes are polyhedral cells that have an oval nucleus, granular chromatin, a single small nucleolus, and a moderate amount of blue-gray cytoplasm (Fig. 17). Macrophages are frequently also present in small numbers. Aspiration of a liver with LSA may also contain increased numbers of lymphocytes. In young ferrets, the cells are usually large immature

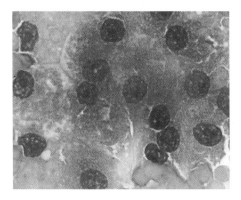

Fig. 17. Normal hepatocytes.

lymphocytes that are never normally present in the liver in any significant numbers. In adult ferrets with the lymphocytic form of LSA, however, the cytologic diagnosis may be difficult to make unless the cells are in very large numbers among the blood cells or are seen in masses interspersed with the hepatocytes distributed in clusters within the smear.

## References

[1] American Ferret Association. The AFA web page. Available at: http://www.ferret.org. Accessed June 13, 2006.
[2] Brown SA. Neoplasia. In: Hillyer EV, Quesenberry KE, editors. Ferrets, rabbits, and rodents: clinical medicine and surgery. Philadelphia: WB Saunders; 1997. p. 99–114.
[3] Li X, Fox JG, Padrid PA. Neoplastic diseases in ferrets: 574 cases (1968–1997). J Am Vet Med Assoc 1998;212:1402–6.
[4] Orcutt C. Dermatologic diseases. In: Hillyer EV, Quesenberry KE, editors. Ferrets, rabbits, and rodents: clinical medicine and surgery. Philadelphia: WB Saunders; 1997. p. 115–25.
[5] Stauber E, Robinette J, Basaraba R, et al. Mast cell tumors in three ferrets. J Am Vet Med Assoc 1990;196:766–7.
[6] Dillberger JE, Altman NH. Neoplasia in ferrets: eleven cases with a review. J Comp Pathol 1989;100:161–76.
[7] Li X, Fox JG. Neoplastic diseases. In: Fox JG, editor. Biology and diseases of the ferret. 2nd edition. Baltimore: Williams & Wilkins; 1998. p. 405–47.
[8] Erdman SE, Brown SA, Kawasaki TA, et al. Clinical and pathologic findings in ferrets with lymphoma: 60 cases (1982–1994). J Am Vet Med Assoc 1996;208:1285–9.
[9] Li X, Fox JG, Erdman SE, et al. Cutaneous lymphoma in a ferret (*Mustela putoris furo*). Vet Pathol 1995;32:55–6.
[10] Rosenbaum MR, Affolter VK, Usborne AL, et al. Cutaneous epitheliotropic lymphoma in a ferret. J Am Vet Med Assoc 1996;209:1441–4.
[11] Fox JG. Bacterial and mycoplasmal diseases. In: Fox JG, editor. Biology and diseases of the ferret. 2nd edition. Baltimore: Williams & Wilkins; 1998. p. 321–54.
[12] Lucas J, Lucas A, Furber H, et al. *Mycobacterium genavense* infection in two aged ferrets with conjunctival lesions. Aust Vet J 2000;78:685–9.
[13] Fox JG. Mycotic diseases. In: Fox JG, editor. Biology and diseases of the ferret. 2nd edition. Baltimore: Williams & Wilkins; 1998. p. 393–403.
[14] Miller PE, Pickett JP. Zygomatic salivary gland mucocele in a ferret. J Am Vet Med Assoc 1989;194:1437–8.
[15] Fox JG. Diseases of the gastrointestinal system. In: Fox JG, editor. Biology and diseases of the ferret. 2nd edition. Baltimore: Williams & Wilkins; 1998. p. 274–5.
[16] Parrott TY, Greiner EC, Parrott JD. *Dirofilaria immitis* infection in three ferrets. J Am Vet Med Assoc 1984;184:582–3.
[17] Moreland AF, Battles AH, Nease JH. Dirofilariasis in a ferret. J Am Vet Med Assoc 1986;188:864.
[18] Rosenthal KL. Respiratory diseases. In: Hillyer EV, Quesenberry KE, editors. Ferrets, rabbits, and rodents: clinical medicine and surgery. Philadelphia: WB Saunders; 1997. p. 81.
[19] Hillyer EV. Part II. Cardiovascular diseases. In: Hillyer EV, Quesenberry KE, editors. Ferrets, rabbits, and rodents: clinical medicine and surgery. Philadelphia: WB Saunders; 1997. p. 72–3.

# Evaluation of Cetacean and Sirenian Cytologic Samples

René A. Varela, MS, VMD[a],*,
Kimberly Schmidt, BS[b], Juli D. Goldstein, DVM[c],
Gregory D. Bossart, VMD, PhD[c]

[a]Ocean Embassy, 6433 Pinecastle Boulevard, Suite 2, Orlando, FL 32809, USA
[b]College of Veterinary Medicine, University of Illinois at Urbana-Champaign, 3505 Veterinary Medicine Basic Sciences Building, 2001 South Lincoln Avenue, Urbana, IL 61802, USA
[c]Division of Marine Mammal Research and Conservation, Center for Coastal Research-Marine Mammal Research and Conservation Program, Harbor Branch Oceanographic Institution, 5600 US 1 North, Fort Pierce, FL 34946, USA

The order Cetacea contains several species that have special appeal to the public. Charismatic species, such as the Atlantic bottlenose dolphin (*Tursiops truncatus*), attract great interest in their welfare, research, and conservation. Many zoos and oceanaria now operate highly sophisticated preventative medicine programs and promote conservation and rescue of stranded animals. Increasing morbidity and mortality among cetaceans in the wild are causes for concern, however. Within their niche, free-ranging bottlenose dolphins are among those cetaceans that are at the top of the marine food chain. In this apex position, they are increasingly exposed to high levels of anthropogenic toxins (eg, polychlorinated biphenyls [PCBs]) in the water as well as the effect of climate change, dwindling fish stocks, and harmful algal blooms [1–3]. In addition, underwater testing of advanced remote detection sonar has recently been thought to be associated with mass mortality events in the Canary Islands [4]. Further, mass mortality has accompanied the emergence of porpoise and dolphin morbilliviruses [5]. Even more recently, stranded bottlenose dolphins in the Indian River Lagoon in Florida have shown evidence of a suite of pathologically significant lesions [6,7].

---

This article was funded by Harbor Branch Oceanographic Institution and the Protect Wild Dolphins License Plate Fund.
* Corresponding author.
E-mail address: rvarela@oceanembassy.com (R.A. Varela).

Species in the order Sirenia (dugongs and manatees), many of which are threatened or endangered, seem less able than cetaceans to seize the attention of the public, resulting in a lower degree of public support [8]. These large herbivorous mammals inhabit tropical and subtropical coastal waters around the world. Manatees are listed as endangered in the Endangered Species Act and vulnerable in The International Union for Conservation of Nature and Natural Resource 2004 Red List of Threatened Species. Emerging diseases associated with mortality in these mammals include viral papillomatosis, brevetoxicosis, and mycobacteriosis [9–11]. Historically, approximately one third of Florida's manatee fatalities are human related, most commonly attributable to watercraft collisions [12,13]. The coastal habitat and slow-moving nature of this species make it difficult to escape imminent danger. Increasingly, legislation stating speed and area restrictions is being introduced in countries in which manatees are found. It is unclear whether these measures are sufficient to ensure the short-term population growth or the long-term survival of the species, however. Cold-stress syndrome has recently emerged as a concern and may be considered another anthropogenic cause of mortality in this species. Exposure to cold-water temperatures (lower than 20°C) can result in immunologic compromise, making this mammal vulnerable to opportunistic infections and unusual lesions [14]. Cold-stress syndrome is an increasing cause for concern, because manatees often reside near warm-water power plants in the winter [15]. Older power plants need to shut down in the near future, which may create a shortage of areas that these mammals have come to depend on for warmth [16]. Conservation programs, such as those conducted by zoos, oceanaria, the Wildlife Trust, and the United States Geological Survey (USGS) Sirenia Project, are producing much needed research as well as developing much needed interventions for disease and management issues on such animals as cetaceans, sirenians, pinnipeds, sea otters, and polar bears [1].

Marine mammals have been described as useful sentinels of ocean health based on evidence that the morbidity of marine mammals reflects that of other smaller organisms with whom they cohabit [1,8,13,15–17]. Thus, conservation of marine mammals should be regarded as important for the whole marine ecosystem and not just for the benefit of the species themselves. Thus, species in the orders Sirenia and Cetacea within human care and conservation programs are in increasing demand of veterinary attention, especially when responding to stranding events and for health assessment studies in the wild. Many of these species are threatened or endangered, and there is considerable public support for the conservation of many charismatic species. Additionally, cogent ecologic arguments exist for their conservation so as to maintain ecosystem health.

A rising demand for veterinary care in cetacean and sirenian species presents an ever-increasing need for diagnostic resources. Obtaining a differential diagnosis in any animal requires the veterinarian to compile as much diagnostic information as possible. A thorough physical examination can reveal visual and palpable abnormalities but fails to assess problems beyond a macroscopic level. It has been shown that disease can manifest itself as cellular abnormalities before producing clinical signs [18]. Cytologic evaluation of body fluids can thus be used as an early indicator of disease, allowing the clinician to assess more clearly irregularities identified during a physical examination. Diagnostic cytologic examinations can also be useful in monitoring cellular changes indicating disease progression or regression. Cytology is thus an integral part of health assessment in veterinary medicine and should be available as a vital tool for clinicians working with any species.

Numerous cytology references are readily available for many domestic and nondomestic species [18–21]. A minimal number of references on marine mammal cytology exist to date, however [22,23]. Additional references should allow for a more complete evaluation of a variety of body fluids, which is needed to allow marine mammal practitioners to establish normative values and determine abnormalities existing in cells or flora when disease is suspected. This article seeks to review the use of cytology in the diagnosis of disease in free-ranging and captive cetaceans and sirenians. These species were chosen because of their popularity in oceanaria, conservation efforts being performed on their behalf, or ready availability of samples and sampling opportunities. The Atlantic bottlenose dolphin (*T truncatus*) and the West Indian manatee (*Trichechus manatus*) are used as examples of cetaceans and sirenians, respectively. A consistent and simple technique is illustrated that allows cytologic samples to be prepared and stored, followed by staining at a later date with minimal effort or expense. Such techniques make this reference useful in situations in which funds, facilities, or time may be a limiting factor in clinical practice or in the field. Cytology is likely to be a valuable, relatively "low-tech," and inexpensive diagnostic tool for veterinarians and biologists and can be used as an adjunct to the clinical examination of a cetacean or sirenian.

## Public health considerations

Because marine mammals are potential carriers of zoonotic pathogens, preventative measures should always be used when collecting or handling their body fluids and secretions. Inhalation of the blowhole sputum aerosol or contamination of cutaneous wounds and abrasions with blood or fecal matter may result in human infections. Although there are few published reports of disease transmission from marine mammals to people, the possibility of zoonotic disease is believed to be significant [24,25]. The paucity of zoonotic infection reports in the literature may, in part, also reflect a recognition failure by physicians or a failure by individuals to seek medical assistance.

Bacteria *Brucella* spp, *Clostridium* spp, *Vibrio* spp, and *Edwardsiella tarda* are highly prevalent in wild cetaceans and may represent a potential health threat to human beings [26]. *Vibrio* spp and *E tarda* have been known to produce gastroenteritis, wound infection, and primary septicemia in people [27]. Exposure to these bacteria is often dependent on the work environment, however. A number of *Brucella* isolates have been found in marine mammals, including the common dolphin (*Delphinus delphis*) [28]. There is at least one report of a laboratory worker contracting clinical infection when handling a marine *Brucella* isolate [29]. The main risk with *Clostridium* spp seems to be wound infection. Mycobacteria have previously been isolated from manatees [11], and the protozoa *Toxoplasma gondii* and *Sarcocystis*-like infections have been found in manatees and other marine mammals [30].

In oceanaria, the health of animals is generally maintained at a high level in contrast to their wild counterparts, which often carry high burdens of parasites and pathogenic microorganisms. Therefore, stranded animals should be assumed to be at a higher risk of carrying zoonotic infection [27]. In addition, the stoic nature of marine mammals suggests that stranded wild animals may harbor disease yet show few clinical signs. The apparent stress of the stranding process may suppress their immune systems, enhancing the proliferation and shedding of pathogens. Thus, the most dangerous scenario for zoonotic transmission may occur when workers come into proximity with animal tissues and fluids [27]. Recommendations for health precautions when working with all marine mammals are as follows [27]: (1) face masks should be worn in any situation when breath is being exhaled in close proximity to the face of a handler, for example, when collecting a sputum sample for respiratory cytology; (2) cuts and abrasions on hands and arms should be covered and gloves worn when blood or other secretions from the animal are handled; and (3) pregnant women, young children, aged persons, and those with chronic or immune-suppressive illness should not be permitted to handle ill marine mammals.

## Restraint of the patient for cytologic sampling

Marine mammals are encountered in institutional care as well as in free-ranging states. A benefit of captive care is that many progressive zoos and oceanaria have developed extensive preventative medicine programs. It is now commonplace for animals in human care to be conditioned for performing specific behaviors for medical procedures. These medical behaviors now form the cornerstone for preventative medicine programs, facilitating sample acquisition for clinical pathology as well as for other diagnostic purposes. For example, trained dolphins present their tail flukes for venipuncture, expel aerosolized respiratory secretions from their blowholes, open their mouths for orogastric tubes to be placed, allow placement of tubes to extract feces, and even allow the placement of urinary catheters for aseptic urine collection. Thus, in facility care, regular cytologic evaluation is simplified, allowing the ongoing population health and preventative medicine program to be monitored and baseline data for specific individuals to be obtained. Such behavioral training is commonplace with captive dolphins across Europe and the United States. Currently, however, health sample training is rare with manatees, with only a few animals in the United States and Mexico trained for such procedures.

In untrained captive animals, confining the animal in a shallow pool and then slowly draining the pool until the animal comes to rest on a padded mat may achieve restraint. Caution must be taken when beaching the animals so that their hydration and body temperature are not adversely affected. Thermometer probes may be useful in monitoring rectal temperature, but the most consistent temperature monitoring can be performed with the use of an infrared thermometer (Raytek, Santa Cruz, California) on various skin surfaces. Constant moistening of the skin is performed with water sprayers or saturated sponges if the animal is out of the water for longer than a few minutes [8,29]. Controlled passive restraint and minimizing excessive stress are critical. A calm bottlenose dolphin weighing 230 kg may be successfully restrained with three to four experienced handlers, but an adult manatee typically weighing 900 kg should have a minimum of four to five experienced people to restrain it [10]. Chemical restraint is documented in both species and can be used when the animal is too fractious to be restrained safely or when the procedure is likely to be noxious [31,32,33]. Techniques to capture wild dolphins and manatees using encircling nets have been described and are frequently used successfully [13,15,16].

**Evaluation of cytologic samples for pathologic significance**

The strength of cytologic evaluation rests with its primary role as an objective description of obtained sample components. An accurate assessment of findings can only occur after detailed and consistent documentation of findings has occurred. Cytologic evaluation involves making a morphologic diagnosis (eg, neoplastic, inflammatory, toxic), followed by a causative agent categorization (eg, bacterial, viral, fungal). The latter may not be possible and may depend on more advanced diagnostic testing (Table 1). Most often, cytologic evaluations are conducted in the context of inflammation, although neoplasia is also being increasingly recognized in these species. Granulomatous inflammation, which is rarely appreciated in cytology, is marked by a predominance of macrophages (histiocytes) and sometimes varying quantities of neutrophils, eosinophils, epithelioid cells, and multinucleate giant cells. Chronic inflammation is typically represented by the presence of lymphocytes (small and large lymphocytes as well as plasma cells) and macrophages. Chronic-active inflammation is marked by the presence of a mixed cellular component of neutrophils, macrophages, and sometimes lymphocytes, plasma cells, and eosinophils. Acute cellular inflammatory processes consist predominantly of neutrophils, which may display subtle morphologic changes important to the diagnosis of specific disease processes [19]. Cytologic changes indicative of neutrophil activation and death

Table 1
Recommended recording sheet for cytologic evaluations

| Sample | General overview | | | Inflammatory cells | | | | | | | | Potential pathogens | | | Other notes |
|---|---|---|---|---|---|---|---|---|---|---|---|---|---|---|---|
| | Morphologic preservation | Cellularity | Noncellular debris | Gastric fluid pH | Inflammation | Neutrophils (band and mature) | Lymphocytes | Macrophages | Eosinophils | Plasma cells | Mast cells | Bacteria | Fungi | Protozoans | Nasitrema |
| No. | 1–3 | 1–3 | 0–3 | | 0–4 | 0–3 | 0–3 | 0–3 | 0–3 | 0–3 | 0–3 | 0–3 | 0–3 | 0–3 | 0–3 |
| 1 | | | | | | | | | | | | | | | | |
| 2 | | | | | | | | | | | | | | | | |
| 3 | | | | | | | | | | | | | | | | |
| Other | | | | | | | | | | | | | | | | |

Although gastric pH is not a cytologic finding, it is included as a reference for the other findings.

*Abbreviations:* cells/hpf, number of cells per high-power field, wherein an hpf is considered to be original magnification ×400.

Key to cytologic constituents includes the following: Morphologic preservation (3 = excellent, 2 = fair, 1 = poor); cellularity (3 = high, 2 = moderate, 1 = low); noncellular debris (3 = abundant, 2 = moderate, 1 = mild, none); inflammatory cells (4 = >15 cells/hpf, 3 = 11–15 cells/hpf, 2 = 6–10 cells/hpf, 1 = 1–5 cells/hpf, 0 = absent); and pathogens (3 = abundant, 2 = moderate, 1 = few, 0 = absent).

Morphologic preservation is defined as the retention of inherent cellular characteristics (eg, crisp nuclear and cell membrane margins), cytoplasmic and organelle detail, normal shape and size, and visualization of distinctive cellular features (eg, brush border of columnar epithelial cells). Cellularity is defined as the concentration of all cells within a low-power field, such that high concentration consists of all cells in the sample contacting each other, where as in instances of low cellularity, few low-power fields contain cells. Noncellular debris is defined as all material present within a sample that is not a cellular component of the host, such as mucus, food debris, environmental algae, or plant material. Potential pathogens are defined as those microorganisms with historical disease-causing capacity, but documenting them alone does not infer pathogenicity of specific isolates. Disease can only be confirmed on comprehensive cytologic diagnosis after all components are integrated.

approximate those of karyolysis, pyknosis, and karyorrhexis. Karyolysis, marked by swelling of nuclear lobes and a decrease in staining intensity, is a sign of rapid cell death and a toxic environment for the cells. In this state, normally rose-purple–colored nuclear lobes of healthy neutrophils with their distinct chromatin pattern become a smooth, pink, homogeneous swollen mass. As karyolysis proceeds, the cell membrane is invariably broken and cytoplasmic contents are expelled. Pyknosis, hypersegmentation, and phagocytosis in the presence of minimal karyolysis collectively indicate a relatively slow progressive change and nontoxic environment. Cytologically, pyknosis is characterized by increased nuclear staining, loss of chromatin pattern, and coalescence of nuclear lobes into a single mass with an intact cell membrane. Although the typical neutrophil of blood consists of a nucleus with four or five lobes, in chronic suppurative lesions attributable to toxin-producing bacteria, degrees of neutrophil hypersegmentation can occur. The resultant pyknosis of these cell nuclei could result in cells appearing to be undergoing karyorrhexis. Many of these cellular features are depicted in neutrophils in Fig. 1). Clearly, hypersegmented neutrophils in exudates are not restricted to bacterial inflammation. Therefore, only phagocytosed bacteria in neutrophils would support inflammation of bacterial origin. In the absence of phagocytosis, all other potential causes of acute inflammation (ie, infectious and noninfectious) should be considered. Because bacterial populations can be heterogeneous or monomorphic, they must be quantified and evaluated in the context of inflammation. A heterogeneous bacterial population (without evidence of phagocytosis) is a more positive sign, which is likely indicative of normal bacterial flora, whereas

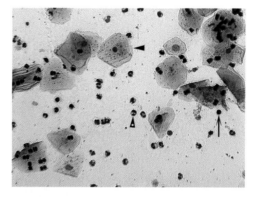

Fig. 1. Sediment from a centrifuged gastric cytologic specimen. Nondegenerate neutrophils (*holed arrowhead*) and lymphocytes (*arrow*) are evident. Partially cornified squamous epithelial cells are present as well (*closed arrowhead*). The inflammation could be of respiratory or gastric origin. A nasal sac cytologic specimen should be examined to determine the origin of the inflammation. Also, a diagnosis cannot be made from a centrifuged sample, because the cells are concentrated. A direct smear should be used for this purpose (*Tursiops truncatus;* modified Wright-Giemsa, original magnification ×200).

a monomorphic population may have pathologic significance. Because yeast organisms, such as *Candida* spp, can be a normal floral component of the upper respiratory, gastrointestinal, and lower urogenital tracts, the diagnosis of a *Candida* infection requires reasoned judgment (eg, quantitative detection of fungal antigens in circulation, presence of inflammation, histologic diagnosis). Other fungal organisms, such as *Aspergillus* spp and zygomycoses, are less often found incidentally in cytologic samples. Observing these organisms (especially reproductive forms, such as budding yeast and septate hyphae) in cytologic samples, even in the absence of inflammation, may be cause for concern and should stimulate a more detailed health evaluation of the animal.

The presence of viruses can also be diagnostically ruled out on cytologic evaluation. Morbilliviruses typically infect cells of the lymphatic, respiratory, intestinal, and urinary systems; the skin; and sometimes the brain, causing syncytial formation and cytoplasmic as well as nuclear inclusion bodies. Herpesvirus can be found in nervous tissue, salivary glands, lymphocytes, and kidneys with evident intranuclear inclusion bodies. Papillomaviruses are found in squamous epithelial cells, rarely displaying basophilic intranuclear inclusion bodies. Influenza viruses typically infect respiratory epithelial, renal, cardiac, and central nervous system cells and produce intranuclear inclusion bodies. Poxvirus infections of the skin may display cytoplasmic inclusion bodies in squamous epithelial cells. A standardized method for evaluating and recording findings of cytologic examinations is recommended and has been used in the subsequent text (see Table 1).

## Respiratory Specimens (sputum)

*Cytologic sampling of the respiratory tract*

The manatee has an upper respiratory tract somewhat similar to that of a terrestrial mammal but with external nares that close completely (Fig. 2).

Fig. 2. Manatee external respiratory tract. (Courtesy of O. Davies, MRCVS, Reading, England, UK).

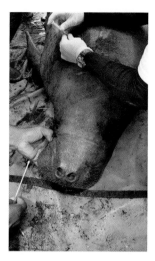

Fig. 3. Collection of a manatee nasal mucosal specimen. (Courtesy of S.D. McCulloch, Fort Pierce, Florida, USA).

A sample of upper respiratory tract secretions can be procured by introducing a sterile microbiologic swab, gently drawing it over the nasal turbinates, and then rolling the swab onto a clean microscope slide (Fig. 3). If the nostril closes while the swab is inserted, it should be withdrawn at the next opening. Removing the swab against a closed nostril may inhibit adequate sample collection.

In cetaceans, the exhaled blowhole sputum can often be diagnostically valuable (Fig. 4). Although cetaceans have bony nasal turbinates and a complex series of nasal sinuses, they possess wide-bore airways through which

Fig. 4. Dolphin respiratory tract cytologic specimens can be collected by using forced expiration. (Courtesy of O. Davies, MRCVS, Reading, England, UK).

exchange of approximately 80% of their lung volume occurs in one breath, in contrast to 20% in human beings. This large gas exchange can result in a rapid expulsion of richly cellular material from the more proximal portions of the respiratory tract that can be simply and noninvasively captured and may be useful to the clinician [23]. Dolphins in human care can be trained to exhale from the blowhole on trainer request, further facilitating the acquisition of a sample for respiratory cytology. In fact, it is our suggestion that the most diagnostically useful samples are obtained by forced exhalation immediately after moderate exercise. A high level of success can be obtained by holding a slide directly in the jet of blowhole sputum. Another method is to place a sterile container in the jet and subsequently transfer the sputum to a slide after a slight dilute with isotonic saline within the collection vial, applying the sputum to a slide with a swab as described previously. In untrained animals, restraint and gentle rocking of the animal from side to side may elicit a forceful chuff from the blowhole [22]. Samples can also be acquired by opportunistically sampling the nasal passage during a breath. If possible, the outer surface of the blowhole should be dried before the procedure is initiated to reduce potential contamination. If results of opportunistic sampling are unsuccessful, respiratory cytology samples can also be obtained by introducing a sterile microbiologic swab into the blowhole when open and gently rubbing the nasal passages (Fig. 5). The blowhole is likely to close with the swab in place; as with the manatee, it should only be removed when the blowhole reopens. Opening of the blowhole may sometimes be induced by gently moving the swab from side to side, stimulating the respiratory mucosa. Use of nonscored culturettes or swabs can help to prevent the swab from breaking while collecting the sample. Gently roll the swab obtained from collection in a thin layer across the surface of a clean histologic-grade microscope slide, and air dry. Make sure to roll the swab across the slide rather than creating a back-and-forth smearing motion, which may distort and rupture the cells. Stain with a modified Wright-Giemsa stain, and air dry. Examine with a microscope under all

Fig. 5. Example of a nasal sac (blowhole) sampling technique with sterile cotton swabs. (Courtesy of S.D. McCulloch, Fort Pierce, Florida, USA).

objective powers. Bronchoalveolar lavage and bronchial brushings have also previously been described in cetaceans and have the potential to yield diagnostic samples from the lower respiratory tract [34,35]. If lower respiratory tract disease is suspected without confirmation on standard cytologic sampling, bronchoalveolar lavage or bronchial brushings via bronchoscopy are strongly recommended.

*Findings in the sirenian respiratory tract*

Common findings in the upper respiratory tract (nares) include squamous epithelial cells, mucus, occasional white blood cells, debris, and small amounts of bacteria (Table 2). Squamous epithelial cells may be used to indicate sample concentration. A variety of bacteria found in the upper respiratory tract of the Antillean manatee have been attributed to the environment in which they reside and are considered normal flora [36]. In the presence of inflammation, however, bacteria may be pathologically significant; indeed phagocytosed bacteria are likely indicative of a bacterial infection if leukocyte numbers are increased. Inflammation in the respiratory tract may be classified on a standard scale (see Table 1). Inflammation should also be evaluated by considering other factors, for example, the number of degenerate and band neutrophils present, large amounts of bacteria, exudate, and high parasite burdens.

Respiratory cytology in the manatee has been poorly documented. The presence of inflammation can be assessed by cytologic methods, however. The normal findings in a healthy manatee include epithelial cells and mucus (Fig. 6) as well as squamous epithelial cells, several (less than five) leukocytes, and a heterogeneous population of bacteria (Fig. 7). In a manatee with purulent nasal discharge, there is evidence of acute inflammation characterized by numerous degranulating reactive manatee neutrophils (with unique, normally occurring, pleomorphic, eosinophilic cytoplasmic granules) and a lymphocyte (Fig. 8). In addition, there is a marked increase in a monomorphic bacterial population and debris. Also, in manatees with purulent nasal discharge, a background of cocci and rod-shaped bacteria, along with multiple degenerating neutrophils, is visible (Fig. 9). Occasionally, parasitic eggs trapped within nasal mucus are noted on cytologic evaluation, but they are likely an incidental finding with no pathologic significance (Fig. 10).

*Findings in the cetacean respiratory tract*

Normal findings in the cetacean upper respiratory tract (blowhole) may include squamous epithelial cells and small amounts of bacteria and algae (Figs. 11–13; see Table 2). Pollen and debris, including fibers from the collection swab, may also normally be found. A small concentration of ciliated protozoans or trematode eggs in the absence of inflammation may be considered normal flora, especially in animals maintained in natural open-water

Table 2
Normal findings and reference ranges for each organ system in cetaceans and sirenians

| Components | Specimen | | | | | |
|---|---|---|---|---|---|---|
| | Respiratory | Gastric | Feces | Urine | Vaginal | Milk |
| Morphologic preservation | Fair–excellent | Fair–excellent | Fair–excellent | Fair–excellent | Fair–excellent | Fair–excellent |
| Cellularity | Low–high | Moderate–high | Moderate–high | Low–moderate | Low–moderate | Low |
| Noncellular debris | Low | Low–moderate | High | Low | Low | Low |
| Inflammation | 0–5 | 0–5 | 0–5 | 0–5 | 0–5 | 0–5 |
| Neutrophils | 0–5 | 0–5 | 0–5 | 0–5 | 0–5 | 0–5 |
| Lymphocytes | 0–5 | 0–5 | 0–5 | 0–5 | 0–5 | 0–5 |
| Macrophage | 0–5 | 0–5 | 0–5 | 0–5 | 0–5 | 0–5 |
| Eosinophils | 0–5 | 0–5 | 0–5 | 0–5 | 0–5 | 0–5 |
| Plasma cells | 0–5 | 0–5 | 0–5 | 0–5 | 0–5 | 0–5 |
| Mast cells | 0–5 | 0–5 | 0–5 | 0–5 | 0–5 | 0–5 |
| Bacteria | Absent–moderate | Absent–moderate | Moderate–abundant | Absent–few | Absent–moderate | Absent–few |
| Fungi | Absent–few | Absent–few | Absent–few | Absent–few | Absent–few | Absent |
| Protozoans | Absent–moderate | Absent–few | Absent–few | Absent | Absent | Absent |
| Nasitrema | Absent–few | Absent–few | Absent–few | Absent | Absent | Absent |

Numeric values are for cells per high-power field.

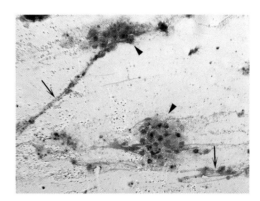

Fig. 6. Mucus strands (*arrows*) and partially cornified squamous epithelial cells (*closed arrowheads*) from the nasal mucosa (*Trichechus manatus*; modified Wright-Giemsa, original magnification ×200).

environments and free ranging. In the presence of numerous leukocytes or erythrocytes, however, such findings may be of pathologic significance. Squamous and columnar epithelial cells are a common finding in a respiratory tract cytology sample and can be used as an indicator of sample concentration. Additionally, a small number of leukocytes (less than five per high-power field) are often present. As an approximate guide to the presence or absence of inflammation, there should be more leukocytes than epithelial cells when inflammation is present [22,23]. Inflammation in the respiratory tract may be classified on a standard scale (see Table 1). Inflammation should also be evaluated by considering other factors, for example, the number of degenerate and band neutrophils present, large amounts of bacteria,

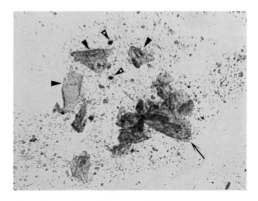

Fig. 7. Cytologic specimen from the nasal mucosa of a manatee demonstrates singular partially cornified squamous epithelial cells (*closed arrowheads*), leukocytes (*holed arrowheads*), and a clump of squamous epithelial cells (*arrow*). A heterogeneous population of bacteria is visible in the background (*Trichechus manatus*; modified Wright-Giemsa, original magnification ×200).

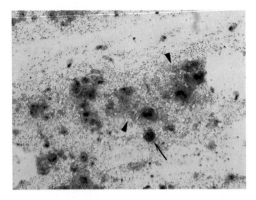

Fig. 8. Typical manatee neutrophils with distinctive eosinophilic granules that have a degranulated appearance (*closed arrowheads*) and a lymphocyte (*arrow*) seen in a nasal swab from a free-ranging manatee (*Trichechus manatus*; modified Wright-Giemsa, original magnification ×400).

exudate, and high parasite burdens. A large number of erythrocytes in the sample are indicative of hemorrhage within the respiratory tract from iatrogenic or disease-associated causes.

Bronchoalveolar lavage samples can be evaluated in a standardized and systematic way [34,35]. Enumeration of leukocytes can be performed within a hemocytometer, and leukocyte differentials (leukocyte types per 500 leukocytes counted) as well as a general survey of the entire sample can be performed on the centrifuged or Cytospin sample on a slide. Since this Cytospin method has gained wide acceptance in human and veterinary medicine, bronchoalveolar lavage results can be quite reliable in assessing disease of focal portions of the lung (Figs. 14–16) [34,35].

Commensal microorganisms that can act as opportunistic pathogens include the bacteria *Simonsiella* spp and the fungi *Candida* spp. The

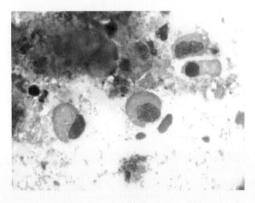

Fig. 9. Typical manatee neutrophils with distinctive eosinophilic granules and amorphous debris from a nasal mucosal cytologic specimen (*Trichechus manatus*; modified Wright-Giemsa, original magnification ×1000).

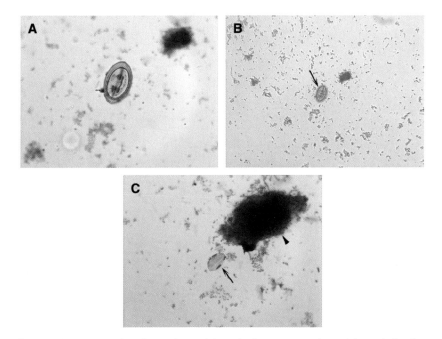

Fig. 10. (*A*) Egg suggestive of protozoan origin and a heterogeneous bacterial population from a nasal mucosal cytologic specimen (*Trichechus manatus*; modified Wright-Giemsa, original magnification ×1000). (*B*) Egg suggestive of protozoan origin (*arrow*) and a heterogeneous bacterial population from a nasal mucosal cytologic specimen (*T manatus*; modified Wright-Giemsa, original magnification ×400). (*C*) Egg suggestive of protozoan origin (*arrow*), amorphous debris (*closed arrowhead*), and a heterogeneous bacterial population from a nasal mucosal cytologic specimen (*T manatus*; modified Wright-Giemsa, original magnification ×1000).

pathogenic significance of these organisms needs to be interpreted in the context of their prevalence, their presence within phagocytic leukocytes, and the degree of inflammation in the sample. If necessary, a definitive diagnosis can be obtained through the histopathologic evaluation of tissue biopsies. Cases of bronchopneumonia may typically exhibit vast amounts of monomorphic bacterial populations, leukocytes, and proteinaceous exudates. Various pathogenic species of fungi may also be encountered. The trematode parasite *Nasitrema attenuata* (Fig. 17) and holotrich ciliate *Kyaroikeus cetarius* (Fig. 18) are frequently found in respiratory cytology samples of the bottlenose dolphin. Adult *N attenuata* occurring in the nasal sinuses and nasal passage of the dolphin are likely nonpathogenic, but eggs located within the bronchial tree may initiate an inflammatory reaction and provoke chronic pneumonia [37]. As with bacteria and fungi, finding these eggs in a sample does not necessarily imply significant disease unless they occur in a large number or in the presence of inflammation. An increased number of ciliates infers a shift in microbial upper respiratory flora and fauna, however, which may signal a subclinical or clinical pathologic state [6].

Fig. 11. Low-power photomicrograph of a moderately cellular nasal sac cytologic specimen with debris (*Tursiops truncatus*; modified Wright-Giemsa, original magnification ×100).

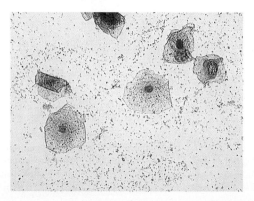

Fig. 12. Heterogeneous bacterial population and multiple partially cornified squamous epithelial cells in a nasal sac cytologic specimen. These are considered normal cytologic findings (*Tursiops truncatus*; modified Wright-Giemsa, original magnification ×400).

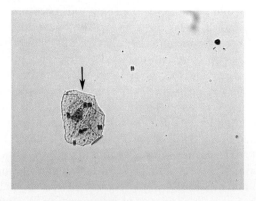

Fig. 13. Normal field from a nasal sac (blowhole) cytologic specimen containing a partially cornified squamous epithelial cell with extracellular algae (*arrow*) (*Tursiops truncatus*; modified Wright-Giemsa, original magnification ×200).

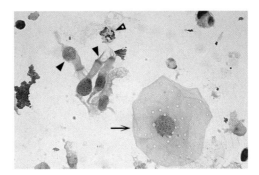

Fig. 14. Bronchoalveolar lavage cytologic specimen shows multiple ciliated columnar epithelial cells (*closed arrowheads*), a partially cornified squamous epithelial cell (*arrow*), and a degenerate neutrophil (*holed arrowhead*) (*Tursiops truncatus*; modified Wright-Giemsa, original magnification ×1000). (Courtesy of F. Townsend, DVM, Fort Walton Beach, Florida, USA).

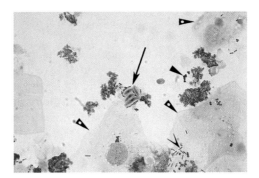

Fig. 15. Bronchoalveolar lavage showing singular squamous epithelial cells (*holed arrowheads*), coccoid bacteria (*closed arrowhead*), rod bacteria (*open arrowhead*), and presumptive protozoans (*arrow*) (*Tursiops truncatus*; modified Wright-Giemsa, original magnification ×1000). (Courtesy of F. Townsend, DVM, Fort Walton Beach, Florida, USA).

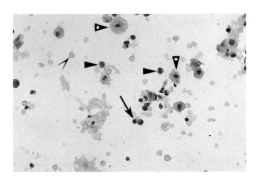

Fig. 16. Cytologic specimen of a bronchoalveolar lavage illustrating macrophages (*holed arrowheads*), neutrophils (*closed arrowheads*), lymphocytes (*arrow*), and red blood cells (*open arrowhead*) (*Tursiops truncatus*; modified Wright-Giemsa, original magnification ×400). (Courtesy of F. Townsend, DVM, Fort Walton Beach, Florida, USA).

Fig. 17. Partially cornified squamous epithelial cells (*closed arrowheads*) and a trematode egg (*Nasitrema* spp; *arrow*) from a nasal sac cytologic specimen. In small numbers and in the absence of inflammation, these parasites are likely a nonpathologic finding. (*Inset*) Magnification of the trematode egg. (*Tursiops truncatus*; modified Wright-Giemsa, original magnifications ×200 and ×400).

Although a common and often insignificant finding, *N attenuata* parasites have been isolated from the respiratory system of several stranded cetacean species [38]. The role that these parasites may play in stranding remains unresolved, however. *N attenuata* can migrate to the brain of a dolphin to cause vestibular neuropathy, encephalitis, and cerebral necrosis [38,39]. In a recent observation by one of the authors (GDB), an adult melon-headed whale, which was part of a mass-stranding event in Florida, was diagnosed histopathologically with a *Nasitrema*-associated, chronic-active, suppurative otitis media. This lesion may have impaired auditory sensation of this single animal, possibly supporting the "follow the leader" theory of mass

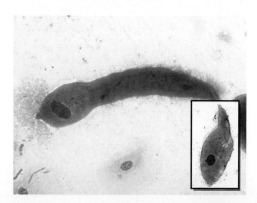

Fig. 18. Partially cornified squamous epithelial cells, large rod bacteria, and a ciliated protozoan, most likely *Kyaroikeus cetarius* (*center field*) from a nasal sac cytologic specimen. (*Inset*) Isolated view of the ciliated protozoan. (*Tursiops truncatus*; modified Wright-Giemsa, original magnifications ×200 and ×400).

stranding events. Larva and eggs from the pathogenic lungworm *Halocercus* spp may also be found in sputum. Parasitic eggs need to be differentiated from pollen particles, diatoms, or algae (Figs. 19–21). Common artifacts in sputum include salt crystals and stain precipitate (Fig. 22).

**Gastric specimens**

*Cytologic sampling of the stomach*

In a manatee, the placement of orogastric tubes can be safely accomplished provided that adequate restraint and lubrication are available. The nasal and oral routes are well tolerated by this species, but the nasal route is more effective [10]. A soft polythene tube that is approximately 2 m long and suitable for a foal or small horse is recommended. It should be premeasured from the most rostral aspect of the animal's face to the caudal aspect of a tucked-in pectoral flipper. The tube should first be covered with an adequate amount of lubricant and then be carefully advanced through the nasal passages or mouth into the pharynx and then into the esophagus. When using the orogastric route, a mouth gag of varying types can be utilized to prevent tube damage from the manatee. Care must be taken to avoid trauma to the caudal oropharynx or accidental intubation of the trachea [10]. When the required tube length has been inserted, placement in the stomach can be ensured by listening down the tube for any gurgling or bubbling sounds. If no sounds are heard, blowing down the tube gently and then listening can often elicit bubbling. In a manatee, fluid is less likely to reflux passively than in a dolphin; thus, gentle suction may be needed to obtain a sample. Once collected, it is often helpful to examine direct and centrifuged samples of the gastric fluid. Fluid is examined directly by dipping a swab into the sample and rolling it onto a slide. A more concentrated

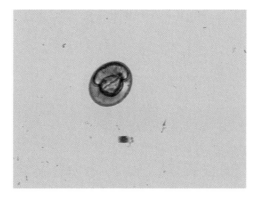

Fig. 19. Organisms suggestive of plant or fungal origin from a nasal sac cytologic specimen (*Tursiops truncatus*; modified Wright-Giemsa, original magnification ×1000).

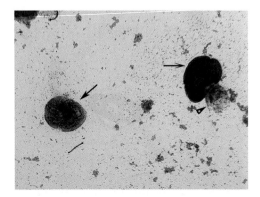

Fig. 20. Nasal sac cytologic specimen shows a partially cornified squamous epithelial cell (*holed arrowhead*) and two unidentified protozoans (*arrows*). In small numbers and in the absence of inflammation, these parasites are a likely a nonpathologic finding (*Tursiops truncatus*; modified Wright-Giemsa, original magnification ×200).

examination of the cellular and solid component of the material is performed by centrifuging the sample, pouring off the supernatant, and then transferring the sediment to a slide. This centrifuged sample is only used to screen for all components, including potential pathogens, within a sample. It is the uncentrifuged direct sample that is used for the primary morphologic diagnosis, because there is no artificial concentration of the sample. Evidence of inflammation in the gastrointestinal tract (gastric and feces) should always be compared with the findings in the respiratory tract. For example, when inflammation in the respiratory tract is severe with a concurrent diagnosis of mild or moderate gastric inflammation, it is possible that the respiratory tract is the primary source of the inflammatory cells or that there is a combined gastric and respiratory inflammatory process.

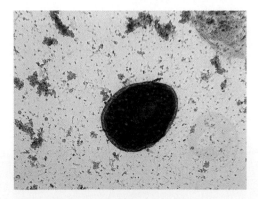

Fig. 21. A partially cornified squamous epithelial cell (*upper right corner*) and unidentified protozoan (*center field*) from a nasal sac cytologic specimen (*Tursiops truncatus*; modified Wright-Giemsa, original magnification ×400).

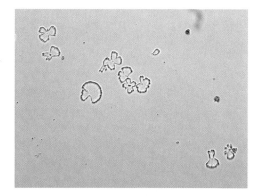

Fig. 22. Artifact of salt crystals in a nasal sac cytologic specimen (*Tursiops truncatus*; modified Wright-Giemsa, original magnification ×400).

As a part of the unique anatomy of cetaceans, the upper respiratory tract is completely separate from the pharynx. The laryngeal lumen is continuous only with a complex network of nasal diverticula, which extend dorsally to the blowhole. As a result, there is no common nasopharynx like that of terrestrial mammals. Thus, gastric tubing a dolphin is often simple, with only a small risk of entering the respiratory tract. Risk of entering the trachea does still exist as a result of the dislocating nature of the dolphin larynx. Dolphins can luxate their larynx cranioventrally, and this allows the laryngeal lumen to communicate with that of the pharynx. As a result of the potential communication between the laryngeal and pharyngeal lumens, cells and secretions from the respiratory tract can be found in the stomach contents.

Dolphins in human care are commonly trained to open their mouth and obligingly swallow an orogastric tube. For stomach tubing wild-captured dolphins, their mouth needs to be held slightly open with the use of towels wrapped around the maxilla and mandibles and their bodies supported so that they do not damage their pectoral fins or injure anyone with their tail flukes. It is recommended to place the maxillary towel first, because the mandible may be prone to fracture if the patient thrashes around under the stress of restraint [22]. It is suggested that gastric contents be collected with an endoscope or a stomach tube that has a maximum diameter of 2 cm. Gastric contents can also be collected with a polyethylene tube. Tubing length required should be measured as the distance between the distal aspect of the animal's rostrum and the cranial edge of the dorsal fin and marked accordingly. Cautious advancement of the tube into the pharynx should then be performed. Resistance is usually encountered from the larynx, bisecting the pharynx in the midline from a ventral to dorsal direction (Fig. 23). When this resistance is encountered, the tube should be rotated slightly while applying light forward pressure; the tube is then guided past the larynx and continues advancing into the esophagus. Access to the

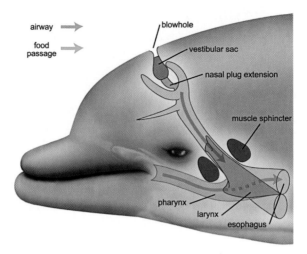

Fig. 23. Diagram of dolphin laryngeal communication with esophagus. (Courtesy of E. Murdoch, BFA, BS, Boca Raton, Florida, USA).

stomach occurs from either side of the larynx, with the tube being passed until the measured length is reached. Correct positioning in the stomach is often confirmed by gastric fluid refluxing spontaneously into the tube. If no fluid refluxes, correct placement of the tube can be determined by listening down the tube for gurgling noises (Fig. 24). Fluid reflux may occur spontaneously when the tube reaches the forestomach. If reflux does not occur, fluid can be evacuated using a syringe to apply gentle suction, and the tube is then slowly removed. Whenever possible, it is advisable to collect gastric samples when the stomach is empty; thus, in facility populations, this collection is best timed before the first feed of the day [22]. Early morning gastric sampling can purposefully assess the significance of undigested food in the stomach. Before cytologic diagnosis, the pH of the sample should always

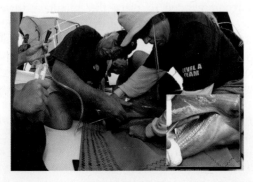

Fig. 24. Gastric sample collection from an Atlantic bottlenose dolphin (*Tursiops truncatus*). (*Inset*) Focus of tube placement location in the mouth.

be measured using pH test strips or, preferably, an electronic pH probe. Low pH of the cetacean stomach (range: 1.5–3.0 pH) has an important effect on the interpretation of the slide. Under normal conditions, the high acidity alters the structure of epithelial cells, lyses erythrocytes, and leaves leukocytes in varying degrees of degradation. Stomach acidity often depletes the cellular cytoplasm, although sparing the nuclei. Under pathologic conditions, the pH is often elevated, resulting in greater preservation of cell morphology. This in itself can be diagnostically informative. For the direct smear, insert a sterile swab into the sample within the tube and then remove it. Gently roll the swab in a thin layer across the surface of a clean histologic-grade microscope slide, and air dry. Make sure to roll the swab across the slide rather than creating a back-and-forth smearing motion, which may severely distort and rupture the cells. Stain with a modified Wright-Giemsa stain, and air dry. Examine with a microscope under low- and high-power fields. For the centrifuged smear, cap the filled tube and place it in centrifuge. Insert a balance tube filled with the same amount of water opposite the sample tube and close the lid. Centrifuge at 400 relative centrifugal force (RCF) or at 1500 to 2000 rpm for 5 minutes. Be cautious not to overcentrifuge the sample. With extremely compacted pellets, resuspension is difficult and cells may compact together, producing an unevenly distributed sample. Remove the tube from the centrifuge after the rotor has come to a stop. Aspirate and discard the supernatant, leaving approximately 0.6 to 1 mL in the tube. Resuspend the gastric sediment by tapping the bottom of the tube. Roll approximately one drop of gastric sediment onto a clean histologic-grade microscope slide in a thin layer using a sterile swab, and air dry. Make sure to roll the swab across the slide rather than creating a back-and-forth smearing motion, which may severely distort and rupture the cells. Stain with a modified Wright-Giemsa stain, and air dry. Examine with a microscope under all objective powers.

Examining the centrifuged sample allows a greater number of cells to be examined for normal morphology, inflammation, neoplasia type, or evidence of dysplasia. A smear from a centrifuged sample can be an extremely useful adjunct to observation of a direct smear, but it should never be used to make an initial diagnosis because it is an amplification of the true concentration in the organ system. Overcentrifugation should be avoided to avert distortion of cellular morphology.

*Findings in the cetacean stomach*

Normal findings in the stomach on cytologic evaluation can include a number of factors (see Table 2). Cetacean stomachs are relatively complex, with a distinct squamous-lined muscular forestomach, glandular fundic chamber, and muscular mucus-producing pyloric chamber. Discrete sphincters separate each chamber, but communication of contents between chambers can occur normally, as with acid reflux from the fundic chamber to the

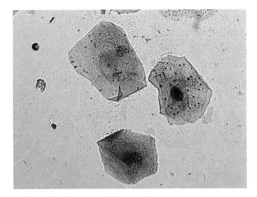

Fig. 25. Partially cornified squamous epithelial cells in a normal dolphin gastric cytologic specimen (*Tursiops truncatus*; modified Wright-Giemsa, original magnification ×400).

forestomach, or pathologically, as with retrograde digesta reflux from the pyloric chamber to the forestomach. Because gastric fluid sampling typically occurs in the forestomach, it is clear that cellular and fluid components of all three chambers may be represented in the sample. Thus, the presence of squamous, cuboidal, columnar, and goblet epithelial cells (respectively) may occur in a gastric sample and can be used as an indicator of concentration (Fig. 25).

Before inflammation can be diagnosed, the presence of leukocytes needs to be considered in respect to their abundance as well as their morphologic type and maturity (Figs. 26–28). In addition, bacteria and fungi may or may not be pathogenically significant (Fig. 29); therefore, their abundance and association with inflammation need to be considered. Erythrocytes are normally lysed by a low pH. Erythrocytes may be visible when the pH has been pathologically raised or when there is profuse hemorrhage into the stomach

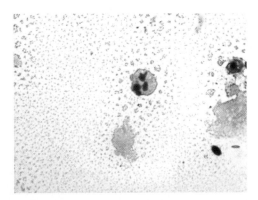

Fig. 26. Degenerate neutrophil in a gastric cytologic specimen, likely attributable to the low pH of the dolphin's stomach (1.5–3 pH) (*Tursiops truncatus*; modified Wright-Giemsa, original magnification ×1000).

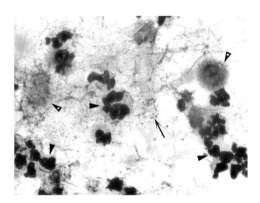

Fig. 27. Neutrophils (*closed arrowheads*), bacteria (*arrow*), and cell debris (*holed arrowhead*) from a gastric cytologic specimen. The neutrophils are well preserved because of a high gastric pH. This photomicrograph represents mild acute inflammation. The inflammation could be of respiratory or gastric origin. A nasal sac cytologic specimen should be examined to determine the origin of the inflammation (*Tursiops truncatus*; modified Wright-Giemsa, original magnification ×1000).

lumen, however. The presence of erythrocytes may indicate hemorrhage or ulceration of the gastric or esophageal mucosa. Care must be taken to determine the cause of hemorrhage or ulceration, because there are several possible causes (eg, *Helicobacter* sp infection, other infectious agent, maldigestion, trauma, nonsteroidal anti-inflammatory drug [NSAID] administration, neoplasia, parasitic migration). Parasites in a gastric sample should be interpreted with caution. Fish in the free-ranging cetacean diet is often inundated with the viable nonpathogenic fish-specific nematode

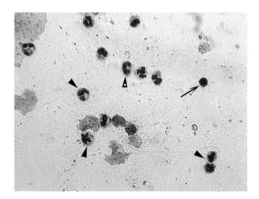

Fig. 28. Mature neutrophils (*closed arrowheads*), a band neutrophil (*holed arrowhead*), and a lymphocyte (*arrow*). A heterogeneous bacterial population is evident in the background. This photomicrograph represents moderate (10–20 cells per high-power field) inflammation. The inflammation could be of respiratory or gastric origin. A nasal sac cytologic specimen should be examined to determine the origin of the inflammation (*Tursiops truncatus*; modified Wright-Giemsa, original magnification ×400).

Fig. 29. Abundant septate, branching, fungal hyphae with parallel walls (*closed arrowheads*) admixed with degenerate squamous epithelial cells and debris in a gastric cytologic specimen (*Tursiops truncatus*; modified Wright-Giemsa, original magnification ×1000).

worms (Fig. 30A). Common gastrointestinal parasites found in the free-ranging cetacean stomach include the nematodes *Anisakis*, *Contracaecum*, and *Pseudoterranova* species [22,38]. Mild infections seldom produce signs, but heavy infections can result in gastritis and ulceration [40,41]. The significance of these organisms in a gastric cytology sample needs to be evaluated with regard to any evidence of inflammation or hemorrhage. In addition, because of the intermittent communication between the laryngeal lumen and the oropharynx of the dolphin, white blood cells and pathogens from the respiratory tract may be found incidentally in the gastrointestinal tract. Trematode parasite eggs *N attenuata* can be found incidentally in gastric cytology samples (see Fig. 30B). The presence of *Nasitrema* sp organisms in the stomach of animals with concurrent trematode encephalitis is significant,

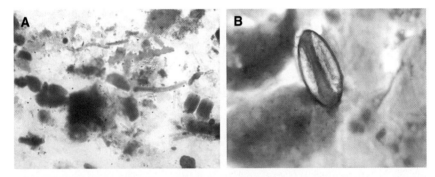

Fig. 30. (*A*) Fish-specific nematode worm (*Anisakis sp*, presumptive) in a dolphin gastric cytologic specimen (*Tursiops truncatus;* modified Wright-Giemsa, original magnification ×1000). (*B*) Egg of the respiratory parasite *Nasitrema* spp found in a gastric cytologic specimen. This egg likely originated in the respiratory tract and was swallowed (*T truncatus*; modified Wright-Giemsa, original magnification ×400).

however. Fish debris and oil droplets are frequent and likely incidental findings in gastric fluid. If the dolphin has been fasted for 12 hours before sample collection, however, the presence of poorly digested fish may indicate maldigestion attributable to a pathologic elevation of stomach pH (Fig. 31). Centrifugation of gastric samples is extremely useful for concentrating the cells in a particularly dilute sample (see Fig. 1).

*Findings in the manatee stomach*

Normal findings on urine cytologic evaluation can include a number of factors (see Table 2). The manatee stomach, simple in comparison to that of cetaceans, contains an unusual accessory digestive gland known as the cardiac gland, which is responsible for producing the acids and mucus necessary for digesting an herbivorous diet. Because manatees are considered monogastric animals, cytologic sampling of their stomachs is likely to yield cellular representatives of each functional area of the stomach. Squamous, columnar, and cuboidal epithelial cells are likely to be found (respectively) and are used to indicate sample concentration (Figs. 32 and 33). The presence of parabasal cells likely indicates healthy gastric mucosa regeneration. Leukocytes may normally be present in small numbers, but inflammation is noted when a high concentration of leukocytes, especially degranulating neutrophils, are present. Small numbers of erythrocytes and basal cells are an occasional, likely iatrogenically caused, finding in normal manatee gastric samples (Fig. 34). The most common pathologic findings in a manatee gastric sample are inflammation and excessive bacteria (Figs. 35–38). Small numbers of fungal hyphae can also be encountered in gastric contents with no apparent pathologic significance (Fig. 39). Inflammation should be classified using the chart included in this article (see Table 1) and considered in the context of bacteria and other pathogens. Other pathogens or plant

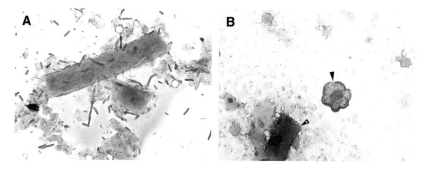

Fig. 31. (*A*) Partially digested food (fish tissue) and large bacilli (likely *Clostridium*) in a gastric cytologic specimen. (*Tursiops truncatus*; modified Wright-Giemsa, original magnification ×1000). (*B*) Swollen fish erythrocyte (*closed arrowhead*) and partially digested fish skeletal muscle (*holed arrowhead*) from a gastric cytologic specimen (*T truncatus*; modified Wright-Giemsa, original magnification ×1000).

Fig. 32. Gastric cytologic specimen with clumps of squamous epithelial cells (*closed arrowheads*), parabasal epithelial cells (*holed arrowheads*), and a heterogeneous bacterial population (*Trichechus manatus*; modified Wright-Giemsa, original magnification ×200).

material may be found in gastric samples. Flagellated myxozoan spores were found incidentally in a few cases (Fig. 40).

## Fecal specimens

### Cytologic sampling of feces

Fecal samples in cetaceans are usually reviewed by a direct smear; however, care must be taken, because the rectal mucosa of a bottlenose dolphin is friable and can be damaged during this procedure. Irritating the rectal mucosa may introduce blood into the specimen that may not have normally been present in the sample. A fecal sample can be obtained by inserting

Fig. 33. Parabasal epithelial cells (*arrow*) and a heterogeneous bacterial population in a gastric cytologic specimen from a manatee (*Trichechus manatus*; modified Wright-Giemsa, original magnification ×200).

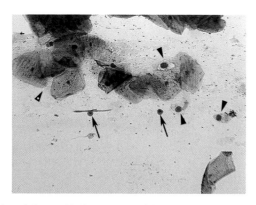

Fig. 34. Clumps of partially cornified squamous epithelial cells (*holed arrowhead*), basal epithelial cells (*closed arrowheads*), and two erythrocytes (*arrows*) are evident in this gastric cytologic specimen. The small number of erythrocytes is likely attributable to trauma of intubation and is nonpathologic (*Trichechus manatus*; modified Wright-Giemsa, original magnification ×200).

a flexible polyethylene tube (maximum diameter of 0.5 cm) through the anal orifice approximately 15 cm into the rectum. A small amount of sterile saline (approximately 5 mL) is flushed through the tube into the rectum, followed by gentle aspiration into a syringe and prompt transfer of the sample to a sterile conical container (Fig. 41). Fecal collection is considerably easier in a conditioned animal that can be instructed to roll into dorsal or lateral recumbency in the water, exposing the anal slit. An observation of inflammation in the feces should be treated as with gastric samples because it may originate from anywhere along the gastrointestinal tract or possibly from the respiratory tract. For this reason, these cytologic samples should be compared with respiratory and gastric specimens.

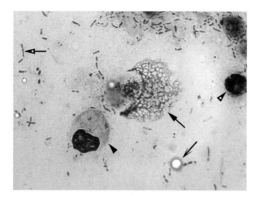

Fig. 35. A macrophage (*closed arrowhead*), a segmented leukocyte (*holed arrowhead*), a smudged epithelial cell (*closed-headed arrow*), large bacilli (*hole-headed arrow*), and a fat droplet (*open-headed arrow*) from a gastric cytologic specimen (*Trichechus manatus*; modified Wright-Giemsa, original magnification ×1000).

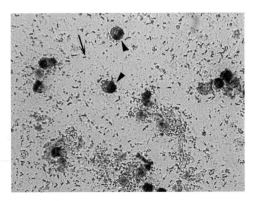

Fig. 36. Primarily neutrophils typical of a manatee (*closed arrowheads*) admixed with a monomorphic population of bacilli (*arrow*), suggesting that the bacteria in this gastric cytologic specimen may be of pathologic significance (*Trichechus manatus*; modified Wright-Giemsa, original magnification ×400).

Manatees are cecocolic fermenters, similar to a horse, and produce large volumes of feces frequently. Thus, feces collection in a manatee is often opportunistic. In a manatee with gastrointestinal hypomotility or ileus, fecal material may be collected as described for a dolphin. Once the sample is collected, it should be smeared directly onto microscope slide. Unlike most domestic species, fecal flotation is not commonly performed in marine mammals. Eggs from most aquatic species do not float in standard concentrations of flotation salt solutions, requiring the less desirable task of specific gravity optimization of flotation solutions for each species [23]. Therefore, a direct smear is preferred for visualizing common parasites in cetaceans

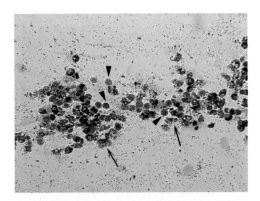

Fig. 37. A gastric cytologic specimen with toxic neutrophils (*arrows*) and lymphocytes (*closed arrowheads*). The eosinophilic granules are a normal finding and unique to the manatee neutrophil. Abundant bacteria are present in the background. The inflammatory cells and toxic changes likely represent severe subacute gastritis of bacterial origin (*Trichechus manatus*; modified Wright-Giemsa, original magnification ×200).

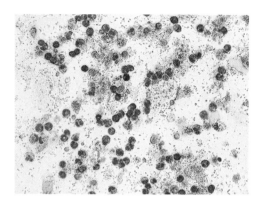

Fig. 38. Highly cellular cytologic specimen from a manatee. The cells are primarily degenerating and nondegenerating neutrophils. The eosinophilic granules are a normal finding and unique to the manatee neutrophil. In the background, a heterogeneous bacterial population is evident. This photomicrograph represents severe acute gastric inflammation (*Trichechus manatus*; modified Wright-Giemsa, original magnification ×400).

and sirenians. To attain a smear, insert a sterile swab into the sample within the tube and remove. Place a drop of sterile saline onto a clean histologic-grade microscope slide. Gently roll swab in a thin layer across the surface of the slide, diluting the sample slightly with the saline. Make sure to roll the swab across the slide rather than creating a back-and-forth smearing motion, which may severely distort and rupture the cells. Stain with a modified Wright-Giemsa stain, and air dry. Examine with a microscope under all objective powers.

*Findings in cetacean feces*

Normal findings in feces (see Table 2) may originate from any location in the gastrointestinal and respiratory tracts. In a normal cetacean fecal

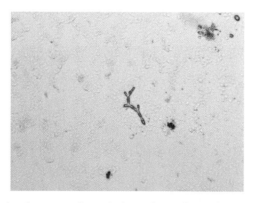

Fig. 39. Fungal hyphae from a gastric cytologic specimen. The small numbers of the organism indicate that it is likely a nonpathologic finding (*Trichechus manatus*; modified Wright-Giemsa, original magnification ×400).

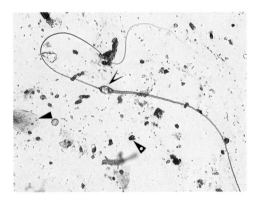

Fig. 40. Gastric cytologic specimen with partially cornified squamous epithelial cells (*closed arrowhead*), neutrophils typical of a manatee (*holed arrowhead*), background bacteria, and a flagellated myxozoan spore (*open arrowhead*). The spore is likely a nonpathologic finding (*Trichechus manatus*; modified Wright-Giemsa, original magnification ×400).

sample, columnar and squamous epithelial cells are present and can be used as an indicator of sample concentration. Epithelial cells in various stages of maturation may also be present (Fig. 42). A monomorphic bacterial population in the presence of numerous leukocytes (greater than five per high-power field) likely represents severe chronic active inflammation of a bacterial cause. This inflammation could be gastrointestinal or respiratory in origin, necessitating analysis of a nasal sac cytologic specimen to determine the origin of the inflammation (Fig. 43). Heterogeneous bacteria populations are often of no clinical significance unless evidence of significant inflammation is found in the feces. Significant pathologic findings that may be associated with inflammation include budding yeasts and numerous parasite eggs and larvae (Fig. 44). Erythrocytes may normally be found as a result of potential mucosal damage caused by the sample collecting tube. A larger number of erythrocytes (greater than five per high-power

Fig. 41. Fecal sample collection from an Atlantic bottlenose dolphin (*Tursiops truncatus*) using sterile saline.

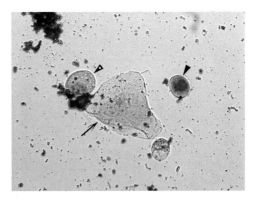

Fig. 42. A dolphin fecal cytologic specimen illustrating squamous epithelial cells in various stages of maturation and background bacteria. There is a basal epithelial cell (*closed arrowhead*), a small, intermediate, noncornified epithelial cell (*holed arrowhead*), and a cornified epithelial cell (*arrow*) (*Tursiops truncatus*; modified Wright-Giemsa, original magnification ×400).

field) may indicate a clinically significant change. Normal findings include partially digested fish material and debris.

*Findings in sirenian feces*

As with cetaceans, normal findings in feces (see Table 2) of sirenians may be representative of any location in the gastrointestinal and respiratory tracts. In a normal sirenian fecal sample, columnar and squamous epithelial cells are present and can be used as indicators of sample concentration (Fig. 45). A high concentration of a mixed population of bacteria is often found (Figs. 46 and 47). Heterogeneous bacterial populations are often of no clinical significance unless evidence of significant inflammation is found

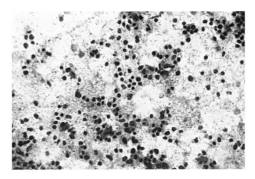

Fig. 43. Numerous leukocytes and a monomorphic bacterial population from a fecal cytologic specimen, which likely represent severe chronic active inflammation of a bacterial cause. This inflammation could be gastrointestinal or respiratory in origin. A nasal sac cytologic specimen should be examined to determine the origin of the inflammation (*Tursiops truncatus*; new methylene blue, original magnification ×1000). (Courtesy of F. Townsend, DVM, Fort Walton Beach, Florida, USA).

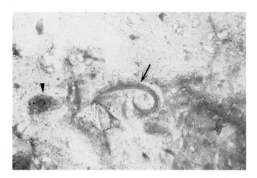

Fig. 44. A partially cornified squamous epithelial cell (*closed arrowhead*) and a helminth parasite (*Halocercus*; *arrow*) found in a fecal cytologic specimen (*Tursiops truncatus*; Gram stain, original magnification ×400). (Courtesy of F. Townsend, DVM, Fort Walton Beach, Florida, USA).

in the feces. Significant pathologic findings that may also be associated with inflammation (greater than five cells per high-power field) include protozoal and helminth parasite eggs and larvae (Figs. 48–50B, C). Erythrocytes may normally be found as a result of potential mucosal damaged caused by the sample collecting tube. A larger number of erythrocytes (greater than five per high-power field) may indicate a clinically significant change. Normal findings include pollen and plant material (see Fig. 50A, D–F; Figs. 51 and 52). Although not demonstrated here, cytologic evaluation is an important diagnostic component in managing cases of neonatal pneumatosis intestinalis and cold-stress–induced colitis in manatees [10,33]. Because diminished immunologic surveillance of the gastrointestinal tract is associated with the disease syndromes in these cases, it is possible that cytologic evaluation may be useful in elucidating a dominant or monomorphic population of a pathogenic organism (ie, bacterial, fungal, protozoal) in the

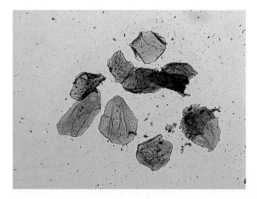

Fig. 45. Partially cornified squamous epithelial cells in a fecal cytologic specimen. This is a normal finding of no pathologic significance (*Trichechus manatus*; modified Wright-Giemsa, original magnification ×200).

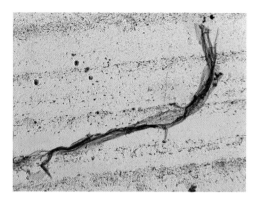

Fig. 46. Debris and abundant bacteria from a fecal cytologic specimen (*Trichechus manatus*; modified Wright-Giemsa, original magnification ×200).

presence of severe inflammation (greater than 15 leukocytes per high-power field). Also, high cellularity comprising squamous, columnar, and goblet epithelial cells may be present in the samples, because blood supply to the colon is compromised. In either case, cytologic evaluation serves as a valuable disease progression monitoring tool.

## Urine specimens

### Cytologic sampling of urine

Urinalysis is perhaps underused as a diagnostic tool in marine mammals. When it is performed, urine is most commonly collected by a free-catch method in the dolphin and manatee. Captive male dolphins can be trained to provide a nonsterile free-catch urine sample by extruding their penis and

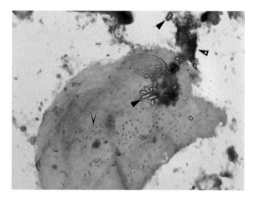

Fig. 47. Cornified squamous epithelial cell with surface bacteria (*open arrowhead*), debris (*holed arrowhead*), and probable bacilli (*closed arrowhead*) from a fecal cytologic specimen (*Trichechus manatus*; modified Wright-Giemsa, original magnification ×1000).

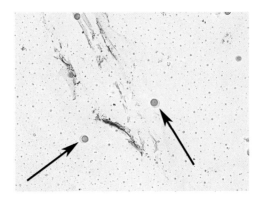

Fig. 48. Organisms suggestive of coccidia (*arrows*) in a manatee fecal cytologic specimen (*Trichechus manatus*; modified Wright-Giemsa, original magnification ×400).

urinating on command. Manatees also can be trained to urinate as a medical behavior; however, such conditioned animals are rare [42]. When sampling by free catch is used, the sample should be collected from the middle of the urine stream. As with all other animals, the recommended method for urine collection is by the aseptic placement of a urinary catheter, which is the standard method of urine collection in wild dolphin health assessment studies and at several oceanaria.

A sterile, lubricated, number 8 French urinary catheter (Mentor Corporation, Santa Barbara, California) can be used to catheterize a dolphin (Fig. 53). Water-soluble lubricants should be used to help prevent trauma to the urethra and bladder while leaving little irritant residue. Urinary catheterization of the manatee is often difficult because of the small tight urethral opening and an often large uncooperative patient [10]. Manual

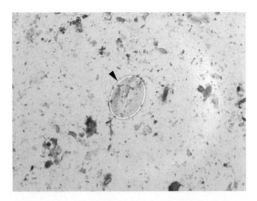

Fig. 49. Organism suggestive of protozoan origin (*closed arrowhead*), amorphous debris, and bacteria present in a fecal cytologic specimen from a manatee (*Trichechus manatus*; modified Wright-Giemsa, original magnification ×1000).

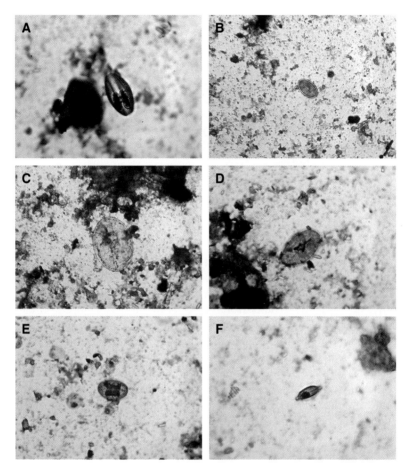

Fig. 50. (*A, D–F*) Organisms suggestive of plant or possibly fungal origin. (*B, C*) Organisms suggestive of protozoal origin from fecal cytologic specimens (*Trichechus manatus*; modified Wright-Giemsa, original magnification ×1000).

stimulation of urine flow by pressing the abdomen cranial to the vulva in female manatees or caudal to the genital opening in male manatees can occasionally meet with some success. The collector should patiently wait for urine to come from the genital slit while a sterile, 15-mL, graduated conical tube is held within the urine midstream. Replace the tube cap when sample collection is complete. Conical bottom test tubes are preferred in this situation because they allow for better pellet formation when centrifuging, similar to gastric sample collection. Cystocentesis, although a common method for urine collection in small domestic animals, is not used in marine mammals, because there are no landmarks and because the inability to palpate the bladder as well as the considerable needle length needed render this procedure obsolete in the face of much safer and efficient methods.

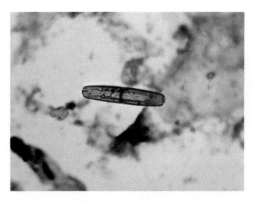

Fig. 51. Organisms suggestive of plant or possibly fungal origin from fecal cytologic specimen (*Trichechus manatus*; modified Wright-Giemsa, original magnification ×1000).

Place the urine-filled sample tube in a centrifuge. It is imperative to centrifuge and examine the sample as soon as possible, because cells left standing at room temperature may undergo degenerative changes attributable to osmotic effect, pH, or toxicity from bacterial toxins [19]. If a sample cannot be examined immediately, it should be refrigerated and brought back to room temperature before examination [22,23]. Centrifuge at 400 RCF or at 1500 to 2000 rpm for 5 minutes, being cautious not to overcentrifuge the sample. With extremely compacted pellets, resuspension is difficult and cells may compact together, producing an unevenly distributed sample. Aspirate and discard the supernatant, leaving approximately 0.6 to 1 mL in the tube. Resuspend urine sediment by tapping the bottom of the tube. Roll approximately one drop of urine sediment onto a clean histologic-grade microscope slide in a thin layer using a sterile swab, and air dry. Make sure to roll the swab across the slide rather than creating a back-and-forth smearing

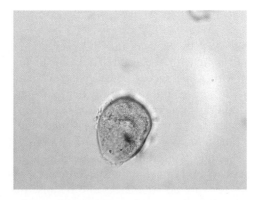

Fig. 52. Pollen in a fecal cytologic specimen (*Trichechus manatus*; modified Wright-Giemsa, original magnification ×1000).

 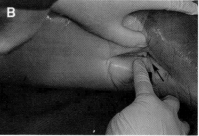

Fig. 53. (A) Catheterization of a male Atlantic bottlenose dolphin (*Tursiops truncatus*) for a urine sample. (B) Catheterization of a female Atlantic bottlenose dolphin (*T truncatus*) for a urine sample. The urethral opening in females (*arrow*) is somewhat less obvious.

motion, which may severely distort and rupture the cells. Stain with a modified Wright-Giemsa stain, and air dry. This type of stain is preferred for the dry mount, because cells appear in better condition and bacteria are more easily visible [19]. Examine with a microscope under all objective powers. This dry mount technique is ideal for the evaluation of cellular morphology and the presence of pathogens. Wet mount preparations are preferred, however, in instances in which the presence of urinary calculi or cellular movement must be determined. These preparations are performed in the same manner, replacing the smearing step with the application of a drop of sediment onto a clean microscope slide with a coverslip, and are then observed under all levels of microscopic power.

*Cytologic findings in cetacean urine*

Normal findings on urine cytologic evaluation can include a number of factors (see Table 2). Urine samples should first be evaluated on low power to detect the presence of large objects, such as crystals, casts, and epithelial cells. A variety of epithelial cells, such as squamous, transitional, and renal epithelial cells, originating throughout the urinary tract may be visualized in a urine sample. The presence of squamous epithelial cells in a sample may be an indication of contamination from the outer skin or genital tract. These cells are most often a result of normal epithelial lining shedding or catheterization (Fig. 54) [18]. Cystitis or other infections of the urogenital tract can result in increased cellularity of the sample. Transitional epithelial cells are present along most of the urinary tract [18]. Transitional epithelial cells from the urethra and bladder are larger and more elliptic than those originating in the ureter and renal pelvis, which are smaller and rounder (Fig. 55). Renal tubular cells are small and round, with a distinct round nucleus. A small number of renal tubular cells in a sample may be a result of normal sloughing. These renal tubular cells may appear in greater concentration if a sample is obtained by catheterization [18]. Goblet cells from the prostate (dolphins only) and cells from the vaginal vault and urethra as well as leukocytes and

Fig. 54. A squamous epithelial cell in a urine cytologic specimen, which likely represents a nonpathologic finding (*Tursiops truncatus*; modified Wright-Giemsa, original magnification ×200).

erythrocytes may also be present in the sample. Occasionally, when urine samples are obtained with a catheter, erythrocytes are present. More than five erythrocytes per high-power field is cause for concern and may be indicative of hemorrhage attributable to excessive mucosal damage of iatrogenic or disease origin. Semen may be seen as well as small amounts of bacteria. Fungal organisms may grow as contaminants but may not be of pathologic significance. As previously mentioned, occasional contamination with semen can occur when catheterizing mature male cetaceans (Fig. 56).

Urine crystals and casts may be observed in the samples (Figs. 57–59). Casts are formed in the lumen of the distal and collecting tubules of the kidney. Because of their tubular origin, casts tend to be long, with parallel sides and a cylindric body shape. Types of casts that may be found in domestic animals include hyaline, granular, cellular, and waxy among others. Because no detailed published reports of urine examination are available for

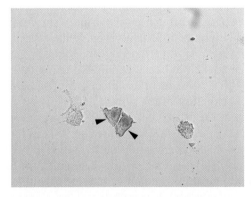

Fig. 55. Transitional epithelial cells (*arrow heads*) in a urine cytologic specimen, which likely represent a nonpathologic finding (*Tursiops truncatus*; modified Wright-Giemsa, original magnification ×200).

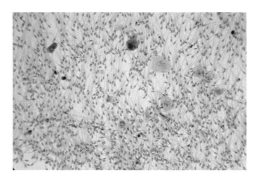

Fig. 56. Squamous epithelial cells and spermatozoa in a male urine cytologic specimen. Spermatozoa may be incidental with no pathologic findings (*Tursiops truncatus*; new methylene blue, original magnification ×400). (Courtesy of F. Townsend, DVM, Fort Walton Beach, Florida, USA).

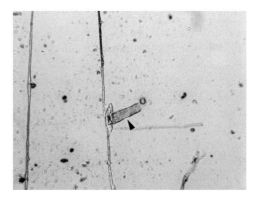

Fig. 57. Urine cast (*arrow head*) from a dolphin, possibly of red blood cell origin (*Tursiops truncatus*; modified Wright-Giemsa, original magnification ×200).

Fig. 58. Unidentified crystal in the urinary sediment of a dolphin. The significance of this crystal is unknown, but no other pathologic findings were present (*Tursiops truncatus*; modified Wright-Giemsa, original magnification ×400).

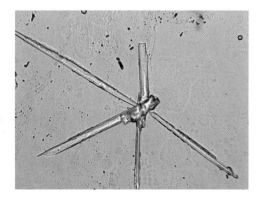

Fig. 59. Unidentified crystal in the urinary sediment of a dolphin. The significance of this crystal is unknown, but no other pathologic findings were present (*Tursiops truncatus*; modified Wright-Giemsa, original magnification ×400).

cetaceans, the relevance of various casts is largely unknown. Figs. 41 and 57 show two possible examples of casts found in centrifuged urine sediment.

Urinary crystals (uroliths) are occasionally observed and are likely formed in relation to urine pH (Figs. 58–59). The normal pH of cetacean urine is typically neutral to slightly acidic (range: 5–6 pH); thus, a deviation from this is postulated to promote the formation of uroliths. Many crystals form as a result of their elements being filtered and concentrated normally in the urine. Some may form because of precipitating disease metabolites that combine with other urine elements, however. Crystals may also form from exogenous factors, such as debris, staining, and salt water [18]. Some unidentified crystals (Fig. 60) can be artifact associated with the dry mounting technique. Using a wet mount for evaluation of urinary calculi is strongly recommended.

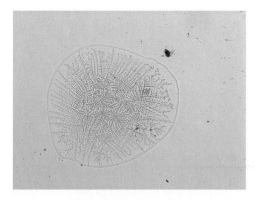

Fig. 60. Artifact: salt crystal in a urine cytologic specimen (*Tursiops truncatus*; modified Wright-Giemsa, original magnification ×200).

## Vaginal specimens

### Cytologic sampling of the vagina

To collect a vaginal cytology swab, the animal should be restrained in lateral recumbency. An experienced technician is needed to retract the folds of the genital slit manually and expose the vaginal opening. A cotton-tipped swab is then passed into the caudal vagina by directing the swab craniodorsal while entering the vaginal vault. Once the swab is cranial to the urethral orifice, it should be gently rotated against the vaginal wall (Fig. 61). Care has to be taken to avoid the vestibule and clitoral fossa so as to avoid sample contamination with superficial cells from these areas, resulting in altered cytologic interpretation. Gently roll the swab obtained from collection in a thin layer across the surface of a clean histologic-grade microscope slide, and air dry. Make sure to roll the swab across the slide rather than creating a back-and-forth smearing motion, which may severely distort and rupture the cells. Stain with a modified Wright-Giemsa stain, and air dry. Examine with a microscope under all objective powers, but evaluate the cytologic components at high power only. The technique for sampling a sirenian vaginal cytology sample is identical to that for a dolphin.

### Findings in cetacean vaginal cytologic specimens

Normal findings on vaginal cytologic evaluation can include a number of factors (see Table 2). A variety of stages of squamous epithelial cell development can be found in a vaginal cytologic sample. Bacteria and leukocytes may also be present to varying degrees. In dogs, these findings are classified and used to indicate the different stages of the reproductive cycle [19]. Cellular findings illustrated here are all analogous to cells found in the canine vagina. Because thorough studies have not yet been published on dolphin vaginal cytologic evaluation, these cells may not yet be classified on their own but can be compared with similar cells in another species. Weekly

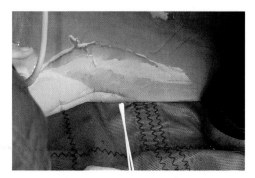

Fig. 61. Vaginal cytologic specimen about to be collected in an Atlantic bottlenose dolphin (*Tursiops truncatus*).

and biweekly blood sampling of trained captive bottlenose dolphins has indicated a seasonality or estrous cycle pattern. Specific estrus cycle stages with associated hormone levels have not been correlated with cellular changes, however [43]. Vaginal cytology is used to monitor the stages of estrous in many domestic species, especially dogs and cats [19]. In these species, proestrus is characterized by a mixture of epithelial cells, particularly parabasal, small and large intermediate cells, and superficial cells. Neutrophils and erythrocytes usually are also present. Estrus cytologic specimens consist mainly of superficial cells. In dogs, superficial cells lose their nuclei, but in cetacean cornified epithelial cells, this step does not occur. Diestrus results in a 20% decrease in superficial cells, and parabasal and intermediate cells increase to greater than 10% and often greater than 50%. Neutrophils and red blood cells appear variably and may be seen in conjunction with epithelial cells. Finally, in anestrus, parabasal and intermediate cells predominate, with bacteria and neutrophils possibly present in small numbers [18].

The characteristics mentioned previously are used in two cases presented here to ascertain the stages of estrous in two bottlenose dolphins. More studies need to be performed to determine the reliability of this system in cetaceans. Hormonal blood assays, in conjunction with cytologic examination, may be extremely useful in determining stages of the estrous cycle in these animals. Also, multiple smears at varying times of the cycle would be useful for monitoring cellular shifts. As an example, animal 1 has a predominance of intermediate epithelial cells, which are small noncornified and large partially cornified (Fig. 62). A small number of neutrophils appear, as do bacteria. No red blood cells are present. Although parabasal cells are absent from the sample, these findings most closely approximate the anestrus stage of the estrous cycle as described previously and may represent a deviation from domestic animal vaginal cytologic findings. Animal 2 has a predominance of parabasal and intermediate noncornified epithelial cells, with an adequate number of bacteria and neutrophils, however (Fig. 63). Some neutrophils are present within the epithelial cells themselves. These findings are characteristic of diestrus in the canine reproductive cycle. In the case of vaginal cytologic examinations, hormonal assays should be performed in conjunction with smears to establish cytologic and hormonal changes concretely throughout the estrous cycle

*Findings in sirenian vaginal cytologies*

Normal findings on urine cytologic evaluation in sirenians can include a number of factors (see Table 2). In manatees, no detailed studies have been published on estrus cycle detection by vaginal cytology. Vaginal cytology samples obtained from two female manatees for this study were characterized by the predominant presence of cornified epithelial cells with a background of a heterogeneous bacterial population (Fig. 64). This cytologic description is indicative of estrus in dogs. Of particular interest is that

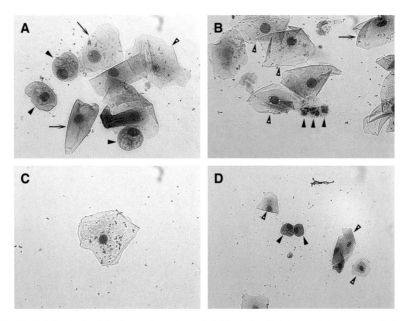

Fig. 62. (*A*) Vaginal cytologic specimen from dolphin 1. Small, intermediate, noncornified epithelial cells (*closed arrowheads*); large, intermediate, partially cornified epithelial cells (*arrows*); and cornified epithelial cells (*holed arrowhead*) (*Tursiops truncatus*; modified Wright-Giemsa, original magnification ×400). (*B*) Vaginal cytologic specimen from dolphin 1. Large, intermediate, partially cornified epithelial cells (*holed arrowheads*); a cornified epithelial cell (*arrow*); and neutrophils (*closed arrowheads*) (*T truncatus*; modified Wright-Giemsa, original magnification ×400). (*C*) Vaginal cytologic specimen from dolphin 1. A large, intermediate, partially cornified epithelial cell with superficial extracellular bacteria. This is a normal finding (*T truncatus*; modified Wright-Giemsa, original magnification ×400). (*D*) Vaginal cytologic specimen from dolphin 1. Large, intermediate, partially cornified epithelial cells (*holed arrowheads*) and small, intermediate, noncornified epithelial cells (*closed arrowheads*) (*T truncatus*; modified Wright-Giemsa, original magnification ×200).

one of these female manatees was lactating at the time of sampling. Although it is generally accepted that marine mammals have lactational anestrus, a resumption of estrus has been observed in cetaceans in response to reduced calf nursing intervals [43]. A return to estrus may indeed be the explanation in this single female manatee. As indicated, vaginal cytologic examination should be performed in conjunction with hormonal assays to establish cytologic and hormonal changes concretely throughout the estrous cycle.

## Milk specimens

### Cytologic sampling of milk

Although it is not a common practice in marine mammal medicine to examine milk samples cytologically, several samples from manatees are

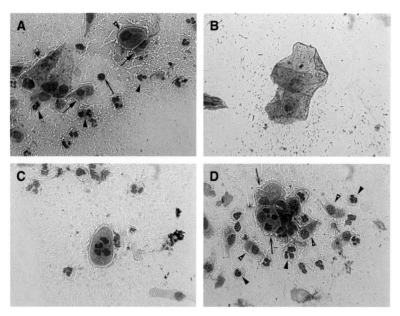

Fig. 63. (*A*) Vaginal cytologic specimen from dolphin 2. Neutrophils (*closed arrowheads*); parabasal noncornified epithelial cells (*closed-headed arrows*); a small, intermediate, noncornified epithelial cell (*holed arrowhead*); and a lymphocyte (*open-headed arrow*). The small numbers of leukocytes in this case likely represent a nonpathologic condition (*Tursiops truncatus*; modified Wright-Giemsa, original magnification ×400). (*B*) Vaginal cytologic specimen from dolphin 2. Large, intermediate, partially cornified epithelial cells with an abundant heterogeneous bacterial population in the background. This probably represents a nonpathologic finding (*T truncatus*; modified Wright-Giemsa, original magnification ×400). (*C*) Vaginal cytologic specimen from dolphin 2. Neutrophils within an epithelial cell (*center field*) (*T truncatus*; modified Wright-Giemsa, original magnification ×400). (*D*) Vaginal cytologic specimen from dolphin 2. Parabasal noncornified epithelial cells (*holed arrowheads*); neutrophils (*closed arrowheads*); and small, intermediate, noncornified epithelial cells, one with enclosed neutrophils (*arrows*) (*T truncatus*; modified Wright-Giemsa, original magnification ×400).

also examined in this article. In domestic animals, cytologic milk evaluation has been quite useful in determining the dam milk constituents and ruling out active mammary inflammation, which may result in poor milk quality [44]. It is likely that such evaluation in marine mammals may produce calf-rearing benefits. Collection is possible in either species by using a 12-mL syringe (plunger removed) placed with the inverted barrel over the nipple within the mammary slits on either side of the genital slit in dolphins or in the axillary region of manatees. A 20-mL syringe should be attached by plastic tubing to the inverted barrel of the 12-mL syringe. Suction should then be applied to the 20-mL syringe to exert gentle pressure on the mammary gland and allow milk to flow into the syringe. The collected sample should then be transferred to a sterile container, and processing of the sample should be performed as described for fluid samples. Insert a sterile swab

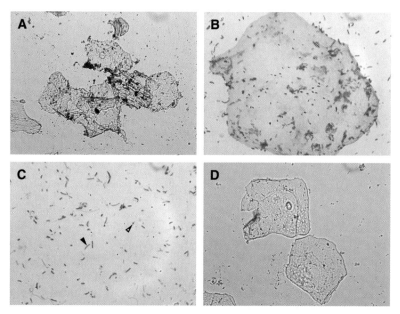

Fig. 64. (*A*) Cornified squamous epithelial cells with debris and background bacteria from a vaginal cytologic specimen. (*Trichechus manatus*; modified Wright-Giemsa, original magnification ×200). (*B*) Cornified squamous epithelial cell with a heterogeneous bacterial population from a vaginal cytologic specimen (*T manatus*; modified Wright-Giemsa, original magnification ×1000). (*C*) Heterogeneous bacterial population in a vaginal cytologic specimen. This represents a nonpathologic finding. Rod (*closed arrowhead*) and coccoid (*holed arrowhead*) bacteria (*T manatus*; modified Wright-Giemsa, original magnification ×1000). (*D*) Cornified epithelial cells and background bacteria in a vaginal cytologic specimen (*T manatus*; modified Wright-Giemsa, original magnification ×1000).

into the sample and remove, gently rolling the swab in a thin layer across the surface of a clean histologic-grade microscope slide, and air dry. Make sure to roll the swab across the slide rather than creating a back-and-forth smearing motion, which may severely distort and rupture the cells. Stain with a modified Wright-Giemsa stain, and air dry. Examine with a microscope under all objective powers.

*Findings in sirenian milk cytology*

Normal findings on urine cytologic evaluation can include a number of factors (see Table 2). In normal milk samples of manatees and most animals, the primary constituent should be nonuniformly sized lipid droplets (Fig. 65). These lipid droplets indicate that the dam is producing high-quality milk with the abundant fat quantity needed for rearing a young calf. Occasionally, in normal milk samples, small amounts of squamous epithelial cells, lymphocytes, or neutrophils can be found (Fig. 66). An increase in any or all of these cellular populations may indicate a pathologic process and

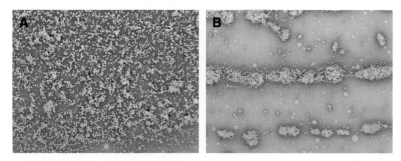

Fig. 65. (*A*) Normal nonuniformly sized lipid droplets from a milk cytologic specimen (*Trichechus manatus*; modified Wright-Giemsa, original magnification ×200). (*B*) Normal nonuniformly sized lipid droplets from a milk cytologic specimen. (*T manatus*; modified Wright-Giemsa, original magnification ×400).

must be evaluated more closely. Mastitis, previously reported in Atlantic white-sided dolphins and beluga whales [45,46], can result in leukocyte elevation and lipid content decrease in milk. Systemic inflammation in the dam, which can cause similar milk composition changes, must be also ruled out diagnostically, however. Primary mammary neoplasia, although not yet

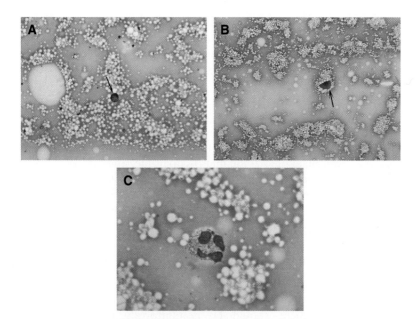

Fig. 66. (*A*) Normal nonuniformly sized lipid droplets and a lymphocyte (*arrow*) from a milk cytologic specimen (*Trichechus manatus*; modified Wright-Giemsa, original magnification ×400). (*B*) Normal nonuniformly sized lipid droplets and a squamous epithelial cell (*arrow*) from a milk cytologic specimen (*T manatus*; modified Wright-Giemsa, original magnification ×200). (*C*) Normal nonuniformly sized lipid droplets and a normal manatee neutrophil from a milk cytologic specimen (*T manatus*; modified Wright-Giemsa, original magnification ×1000).

reported in manatees and dolphins, could result in elevations of milk squamous epithelial cells, lymphocytes, macrophages, and malignantly transformed cells.

**Additional specimens**

When necessary, fine-needle aspirates of tissues and masses, impression smears of open tissues and lesions, scrapings of open lesions (ie, conjunctiva, nonhealing wounds), fluid from abdominocentesis and thoracocentesis, and cerebrospinal fluid (CSF) can be evaluated cytologically in sirenians and cetaceans. These sampling techniques and cytologic interpretations for these samples in sirenians and cetaceans are essentially identical to those established in domestic animals. Modifications must be made to adjust for anatomic differences observed in these species. Also, the highly invasive nature of acquiring samples, such as CSF, from these species must be balanced with the potential benefits of diagnosing the focal central nervous disease. For this reason, sampling of CSF is typically reserved as a postmortem procedure. All the other procedures presented here are of excellent diagnostic potential, which can yield valuable information when used by marine mammal clinicians.

**Summary**

The normal and pathologic findings reported in this article can serve to provide a reference for the future analysis of cytologic samples from cetaceans and sirenians. There is clear clinical usefulness to this information when it is used to diagnose disease and track the response to medical intervention in these species. Because urine sediment, vaginal, and milk cytologic evaluations have rarely been performed in these species, minimal data exist on what is normal or pathologic. Further study is needed before valid reference ranges can be established for cetaceans and sirenians. Also, more information is needed to correlate morphologic and causative diagnoses. Marine mammal medicine is only now beginning to expand the database of causative agents. Our hope is that this article serves as a clinically informative diagnostic tool for clinicians. Formulating a preliminary diagnosis from cytologic changes may allow early disease detection in marine mammals, facilitating rapid medical action and assisting veterinarians in their clinical practice.

**Acknowledgments**

The authors thank Dr. Owen Davies for his contributions to the early portions of this project. They recognize Drs. Roberto Sanchez, Alfonso Lopez, Sandra Novoa, and Jaime Bernal (Dolphin Discovery); Dr. Forrest Townsend (Bayside Hospital for Animals); and Dr. Maya Rodriguez

(Miami Seaquarium) for their generous sample contributions. They also thank Stephen D. McCulloch for his budgetary and personal support and Eric Reece as well as Nelson Beaman for their technical assistance on this project. Finally, the authors thank Dr. R.H. Defran for his conscientious review of this manuscript.

## References

[1] Bossart GD. Marine mammals as sentinel species for oceans and human health. Oceanography 2006;19(2):44–7.
[2] Fleming LE, Backer LC, Baden DG. Overview of aerosolized Florida red tide toxins: exposures and effects. Environ Health Perspect 2005;113(5):618–20.
[3] Jepson PD, Bennett PM, Deaville R, et al. Relationships between polychlorinated biphenyls and health status in harbor porpoises (Phocoena phocoena) stranded in the United Kingdom. Environ Toxicol Chem 2005;24(1):238–48.
[4] Jepson PD, Arbelo M, Deaville R, et al. Gas-bubble lesions in stranded cetaceans. Nature 2003;425:575–6.
[5] Taubenberger JK, Tsai M, Krafft AE, et al. Two morbilliviruses implicated in bottlenose dolphin epizootics. Emerg Infect Dis 1996;2:213–61.
[6] Bossart G, Meisner R, Varela R, et al. Pathologic findings in stranded Atlantic bottlenose dolphins (Tursiops truncatus) from the Indian River Lagoon, Florida. The Florida Scientist 2003;66(3):226–38.
[7] Bossart GD, Ghim S, Rehtanz M, et al. Orogenital neoplasia in Atlantic bottlenose dolphins (Tursiops truncatus). Aquatic Mammals 2005;31(4):473–80.
[8] Bossart GD. The Florida manatee: on the verge of extinction? J Am Vet Med Assoc 1999;214:10–5.
[9] Bossart GD, Ewing RY, Lowe M, et al. Viral papillomatosis in Florida manatees (Trichechus manatus latirostris). Exp Mol Pathol 2002;72:37–48.
[10] Bossart GD. Manatees. In: Gulland F, Dierauf L, editors. The CRC handbook of marine mammal medicine. 2nd edition. Boca Raton (FL): CRC Press; 2001. p. 939–60.
[11] Sato T, Shibuya H, Ohba S, et al. Mycobacteriosis in two captive Florida manatees (Trichechus manatus latirostris). J Zoo Wildl Med 2003;34(2):184–8.
[12] Bossart GD, Meisner R, Rommel SA, et al. Pathologic findings in Florida manatees (Trichechus manatus latirostris). Aquatic Mammals 2004;30(3):434–40.
[13] US Marine Mammal Commission. Annual report to Congress. Bethesda (MD): US Marine Mammal Commission; 2003. p. 52.
[14] Bossart GD, Meisner RA, Rommel SA, et al. Pathological features of the Florida manatee cold stress syndrome. Aquatic Mammals 2003;29(1):9–17.
[15] US Marine Mammal Commission. Annual report to Congress. Bethesda (MD): US Marine Mammal Commission; 2001. p. 114–29.
[16] US Fish and Wildlife Service. Florida manatee recovery plan. 3rd revision. Atlanta (GA): US Fish and Wildlife Service; 2001. p. 1–109.
[17] Reddy ML, Gulland FMD, Dierauf LA. Marine mammals as sentinels of ocean health. In: Gulland F, Dierauf L, editors. The CRC handbook of marine mammal medicine. 2nd edition. Boca Raton (FL): CRC Press; 2001. p. 3–13.
[18] Cowell RL, Tyler RD, Meinkoth JH. Diagnostic cytology and hematology of the dog and cat. St. Louis (MO): Mosby; 1999. p. 338.
[19] Perman V, Alsaker AD, Riis RC. Cytology of the dog and cat. South Bend (IN): American Animal Hospital Association; 1979.
[20] Boon ME, Drijver JS. Routine cytological staining techniques: theoretical background and practice. New York: Elsevier; 1986. p. 238.

[21] Woronzoff-Dashkoff KK. The Wright-Giemsa stain, secrets revealed. Clin Lab Med 2002; 22(1):15–23.
[22] Sweeney JC, Reddy ML. Cetacean cytology. In: Gulland F, Dierauf L, editors. The CRC handbook of marine mammal medicine. 2nd edition. Boca Raton (FL): CRC Press; 2001. p. 437–48.
[23] Sweeney J, Reddy M. Handbook of cetacean cytology. Waikoloa (HI): Dolphin Quest; 2003.
[24] Tryland M. Zoonoses of arctic marine mammals. Infectious Disease Review 2000;2(2):55–64.
[25] Mazet J, Hunt T, Ziccardi M. Assessment of the risk of zoonotic disease transmission to marine mammal workers and the public: survey of occupational risks. United States Marine Mammal Commission. Final report: research agreement number K005486–01. Wildlife Health Center, School of Veterinary Medicine, University of California, Davis; Davis, CA. 2004. p. 1–45.
[26] Cowan DF, Turnbull BS, Haubold EM. Organisms cultured from stranded cetaceans: implications for rehabilitation and for safety of handlers. In: Walsh M, editor. Proceedings of the 29th International Association for Aquatic Animal Medicine. East Falmouth (MA): IAAAM; 1998;30:148–9.
[27] Cowan DF, House C, House JA. Public health. In: Gulland F, Dierauf L, editors. The CRC handbook of marine mammal medicine. 2nd edition. Boca Raton (FL): CRC Press; 2001. p. 767–78.
[28] Ross HM, Jahans KL, MacMillan AP, et al. Brucella species infection in North Sea seal and cetacean populations. Vet Rec 1996;138:647–8.
[29] Brew SD, Perrett LL, Stack JA, et al. Human exposure to Brucella recovered from a sea mammal. Vet Rec 1999;144:483.
[30] Dubey JP, Zarnke R, Thomas NJ, et al. Toxoplasma gondii, Neospora caninum, Sarcocystis neurona, and Sarcocystis canis-like infections in marine mammals. Vet Parasitol 2003;116: 275–96.
[31] Barnett J, Knight A, Stevens M. British Divers Marine Life Rescue marine mammal medic handbook. 5th edition. London: British Divers Marine Life Rescue Society; 2006.
[32] Joseph B, Cornell L. The use of meperidine hydrochloride for chemical restraint in certain Cetaceans and Pinnipeds. J Wildl Dis 1988;24:691–4.
[33] Walsh MT, Bossart GD. Manatee medicine. In: Fowler ME, Miller RE, editors. Zoo and wild animal medicine: current therapy. 4th edition. Philadelphia: WB Saunders; 1999. p. 507–16.
[34] Tsang KW, Kinoshita1 R, Rouke N, et al. Bronchoscopy of cetaceans. J Wildl Dis 2002; 38(1):224–7.
[35] Dover SR, Van Bonn W. Flexible and rigid endoscopy in marine mammals. In: Gulland F, Dierauf L, editors. The CRC handbook of marine mammal medicine. 2nd edition. Boca Raton (FL): CRC Press; 2001. p. 621–42.
[36] Vergara-Parente JE, Sidrim JJC, Pessoa AGPE, et al. Bacterial flora of upper respiratory tract of captive Antillean manatees. Aquatic Mammals 2003;29(1):124–30.
[37] Kumar V, Vercruysse J, Kageruka P, et al. Nasitrema attenuata (Trematoda) infection of Tursiops truncatus and its potentialities as an etiological agent of chronic pulmonary lesions. J Helminthol 1975;49(4):289–92.
[38] Dailey MD, Walker WA. Parasitism as a factor in single strandings of southern California cetaceans. J Parasitol 1978;64:593–6.
[39] O'Shea TJ, Homer BL, Greiner EC, et al. Nasitrema sp.-associated encephalitis in a striped dolphin (Stenella coeruleoalba) stranded in the Gulf of Mexico. J Wildl Dis 1991;27(4): 706–9.
[40] Smith JW. Ulcers associated with larval *Anisakis simplex B* (Nematoda: Ascaridoidea) in the forestomach of harbour porpoises Phocoena phocoena. Can J Zool 1989;67:2270–6.
[41] Matthews BE. The source, release and specificity of proteolytic enzyme activity produced by Anisakis simplex larvae (Nematode: Ascaridida) in vitro. J Helminthol 1984;58:175–85.
[42] Colbert DE, Fellner W, Bauer GB, et al. Husbandry and research training of two Florida manatees (Trichechus manatus latirostris). Aquatic Mammals 2001;27(1):16–23.

[43] Robeck TR, Atkinson SKC, Brook Reproduction F. In: Gulland F, Dierauf L, editors. The CRC handbook of marine mammal medicine. 2nd edition. Boca Raton (FL): CRC Press; 2001. p. 193–236.

[44] Jones GM, Pearson RE, Clabaugh GA, et al. Relationships between somatic cell counts and milk production. J Dairy Sci 1984;67(8):1823–31.

[45] Geraci JR, Dailey MD, St. Aubin DJ. Parasitic mastitis in the Atlantic white-sided dolphin, Lagenorhynchus acutus, as a probable factor in herd productivity. Journal of the Fisheries Research Board of Canada 1978;35(10):1350–5.

[46] DeGuise S, Lagace A, Beland P, et al. Non-neoplastic lesions in beluga whales (Delphinapterus leucas) and other marine mammals from the St. Lawrence Estuary. J Comp Pathol 1995;112(3):257–71.

VETERINARY
CLINICS
Exotic Animal Practice

# Avian Cytology

Kenneth S. Latimer, DVM, PhD, DACVP[a],*,
Pauline M. Rakich, DVM, PhD, DACVP[b]

[a]*Clinical Pathology Laboratory, Department of Pathology,
College of Veterinary Medicine, University of Georgia, 501 D,
West Brooks Drive, Athens, GA 30602, USA*
[b]*Athens Veterinary Diagnostic Laboratory, College of Veterinary Medicine,
University of Georgia, 501 D, West Brooks Drive, Athens, GA 30602, USA*

Cytologic techniques to obtain and prepare avian specimens for microscopic examination are similar to those used for mammalian specimens. Techniques to obtain cytologic specimens include fine-needle aspiration, touch imprints, scrapings, and lavage. If fluid specimens are turbid, direct smears can be prepared. If fluid specimens lack turbidity, concentration techniques should be used to increase the cellular yield. Although most avian cytologic specimens are stained with Romanowsky's or new methylene blue stains, other stains may occasionally be of benefit to demonstrate pathogens and iron. These stains, used primarily by diagnostic pathologists, include Gram stain for bacteria, acid-fast stain for mycobacteria, periodic acid–Schiff and Gomori's methenamine silver staining for fungi, and Perls' stain for iron. In specimens from some individuals, heterophils and basophils may experience stain-associated dissolution of granule matrix when using rapid Romanowsky's stains, such as Diff-Quik. This problem usually can be avoided by extending the fixation time, if necessary.

This article is not exhaustive in scope, but provides a basic, practice-oriented overview of avian cytology. Cytologic specimens usually are classified as inflammatory, infectious, or neoplastic and a similar approach is used in this article. Specific anatomic areas are mentioned where cytologic specimens are frequently obtained in clinical practice. Other major reference sources on this topic also have been published and can be consulted for additional information concerning the preparation, staining, and cytologic evaluation of avian specimens [1–3].

---

* Corresponding author.
*E-mail address:* latimer@vet.uga.edu (K.S. Latimer).

## Inflammation

Inflammation is a component of many cytologic preparations and recognition of these cells is an important step in accurate cytologic diagnosis. Furthermore, the inflammatory cell pattern may alert the microscopist to the presence of specific pathogens.

Heterophilic inflammation is characterized by an abundance of heterophils (Fig. 1A). Heterophils are the avian equivalent of mammalian neutrophils and are involved in bacterial killing [4–6]. The heterophils of most avian species have red to red-brown, needle-shaped, cytoplasmic granules that may partially obscure nuclear morphology. Their cytoplasm is colorless. Heterophils are usually the first inflammatory cells to emigrate from the blood into damaged tissue [7]. Heterophilic inflammation may be associated with acute trauma or bacterial infection. Its presence may incite subsequent granuloma formation.

Eosinophilic infiltrates are less frequent in birds than in mammals but may be observed with parasitism or hypersensitivity (Fig. 1B) [8,9]. Avian eosinophils have red to red-orange, round to irregular granules. The

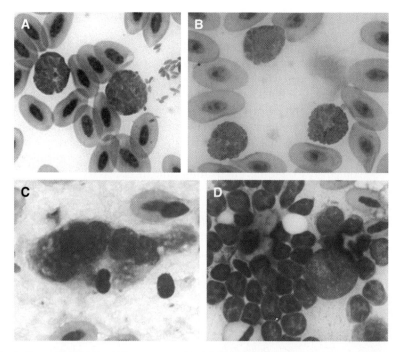

Fig. 1. Examples of avian inflammatory cells. (*A*) Heterophils with pointed, red-brown granules. Notice individual granules at right from ruptured heterophil. (*B*) Eosinophils with round, bright red granules. (*C*) Multinucleated giant cell. (*D*) Small lymphocytes and a plasma cell (Wright's stain).

cytoplasm of eosinophils may have a slight blue cast. Acute hypersensitivity reactions may involve basophils, mast cells, and eosinophils.

Granulomatous infiltrates are composed of macrophages and occasional multinucleated giant cells (Fig. 1C). Degranulation of heterophils often precipitates granulomatous inflammation at a later point in time [10]. In addition, granulomatous inflammation may be prominent in fungal infection; certain bacterial infections (especially avian mycobacteriosis and nocardiosis); and in response to foreign bodies. Macrophages have a round to oval nucleus with abundant gray cytoplasm that is often vacuolated. Most macrophages within lesions are recruited from blood monocytes that further mature in the tissues. Macrophage activation occurs within 4 hours of the onset of inflammation [11]. In some lesions, the macrophages may be large and resemble epithelial cells. Such cells are called "epithelioid macrophages." Multinucleated giant cells are formed by the fusion of macrophages and have multiple round nuclei and abundant gray cytoplasm [12].

Lymphocytic and lymphoplasmacytic infiltrates are usually associated with antigenic stimulation but also may be a feature of delayed-type hypersensitivity (Fig. 1D) [13]. B lymphocytes and T lymphocytes function in humoral and cell-mediated immunity, respectively. In Romanowsky's-stained preparations, lymphocytes may be variable in size and usually have a round to slightly indented nucleus, an aggregated chromatin pattern, and blue cytoplasm. Reactive lymphocytes may have intensely blue cytoplasm or demonstrate plasmacytoid differentiation. Plasma cells are the ultimate expression of B lymphocytes and are responsible for antibody production. Plasma cells appear slightly blocky or rectangular with an eccentrically placed nucleus that has a coarse chromatin pattern. The cytoplasm is moderate in amount, dark blue, and occasionally contains a lighter patch, which represents the Golgi's area that contains immunoglobulin.

Other cells that may be observed in avian inflammatory lesions include fibroblasts, basophils, and mast cells. Fibroblasts indicate chronicity of the lesions. These cells are fusiform and have an oval nucleus; moderately granular chromatin pattern; small nucleoli; and wispy, blue cytoplasm. Avian basophils have round purple cytoplasmic granules and clear cytoplasm [13]. The nucleus generally is not lobated. Although birds other than raptors usually have more basophils than eosinophils in circulating blood, basophils are infrequently observed in avian inflammatory lesions. This may partially be the result of the application of rapid Romanowsky's-type stains that may cause granule dissolution. Mast cells are rare in routine avian cytologic preparations, but have been observed in mast cell tumors of owls and chickens [14]. These cells, when observed, have a round nucleus and purple cytoplasmic granules that may obscure nuclear morphology. Diff-Quik staining may cause dissolution of mast cell and basophil granules. Because the fixation time is so short, the granule matrix may not be adequately stabilized for dye uptake. As the cytoplasmic granules dissolve during staining, the cells may have clear, vacuolated cytoplasm. In some preparations, basophils or mast cells may be surrounded by a purple smudge or "strings" that indicate dissolution of the granule matrix.

## Infection

Infectious agents of birds are as diverse as those of mammals and include bacteria, fungi, parasites, and viruses. General remarks concerning these classes of pathogens follow. Selected pathogens are demonstrated in subsequent sections of this article.

As a general rule, bacteria usually stain dark blue with Romanowsky's stains (Fig. 2A). Common bacterial pathogens in birds include *Staphylococcus* sp (large cocci in clusters); *Streptococcus* sp (smaller cocci in short chains); *Clostridium* sp (large bacilli with spore formation); and *Escherichia coli*, *Salmonella* sp, and *Pasteurella* sp (small bacilli). Mycobacteria are a notable exception to this generalization; they fail to stain because of the high lipid content of the bacterial cell wall (Fig. 2B). These organisms appear as refractile, unstained rods in Romanowsky's-stained preparations but can be specifically stained red by a variety of acid-fast techniques (eg, Ziehl-Neelsen

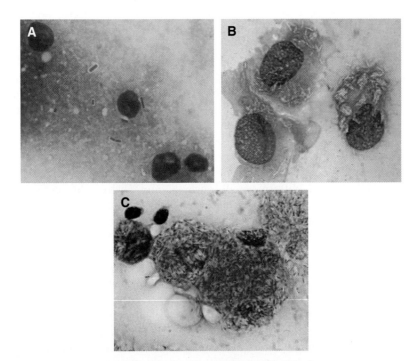

Fig. 2. (*A*) Encapsulated bacilli are present in an hepatic imprint from a bird with sepsis; Macaw chick (Wright's stain). (*B*) Macrophages with unstained bacilli of *Mycobacterium avium* complex (Wright's stain). (*C*) Mycobacteria are red after acid-fast staining (Kinyoun's technique).

technique, Kinyoun technique) (Fig. 2C). Gram stain is often used to demonstrate bacteria, but Gram staining characteristics may be altered in some exudates. Specifically, gram-negative organisms may stain blue if they are senescent or the cell wall has been damaged by inflammatory cell proteases. In addition, overdecolorization of the preparation before counterstaining with safranine dye causes all bacteria to appear red or gram-negative. Gram staining should be viewed as an artistic application of a scientific procedure. Proper attention to technique is necessary for optimal staining results. Furthermore, tinctorial characteristics of bacteria may change with senescence, damage, or overdecolorization. As a note of caution, stain precipitate can interfere with microscopic evaluation of cytologic preparations. Stain precipitate is blue-black and granular and can be mistaken for bacteria. In large enough quantities, stain precipitate can obscure bacteria or cellular elements.

Fungal hyphae and yeasts can be readily observed in Romanowsky's-stained preparations. If staining is faint or the identity of the organism is uncertain, periodic acid–Schiff or Gomori's methenamine silver staining may enhance identification of the offending organism.

Parasites are diverse but may be identified in cytologic preparations by the examination of wet mounts, zinc sulfate flotations, and Romanowsky's– or new methylene blue–stained preparations. Wet mounts of intestinal contents may facilitate the diagnosis of giardiasis, hexamitiasis, and trichomoniasis. Parasitic forms observed cytologically include merozoites, tachyzoites, shizonts, gametocytes, flagellates, sporozoites, sporocysts, and microfilariae [14]. Microfilariae are more commonly observed in cytologic specimens from free ranging songbirds and wild-caught cockatoos (Fig. 3A).

Identification of viral inclusions in cytologic specimens is serendipitous. Aviadenovirus and avian polyomavirus inclusions may be associated with moderate to marked karyomegaly. Circoviral infections are more apparent as multiple cytoplasmic globules in macrophages. More definitive diagnosis of viral inclusions in cytologic specimens may require fluorescent antibody staining, immunocytochemistry, or nucleic acid in situ hybridization.

## Neoplasia

Although the histology of avian neoplasms has been covered in depth [15], the cytologic aspects of neoplasia have received much less attention. Cytologically, the diagnosis of neoplasia progresses through three steps. The first step is to determine whether the lesion is neoplastic. This conclusion is usually based on the observation that the cytologic components of the lesion are not normal for the site sampled; are not characteristic of inflammation; and are composed of a distinct, homogeneous cell population. The second step is to determine if the neoplasm is benign or malignant. A benign neoplasm usually has a very uniform cell population with mature

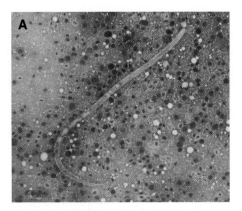

Fig. 3. (*A*) Faintly stained microfilaria in a liver aspirate from a wild-caught cockatoo (Wright's stain). (*B*) Appearance of *Knemidocoptes* sp female (left) and male (right) mites from a psittaciform bird with scaly leg. Mature female mites are much larger than mature male mites.

nuclei and little mitotic activity. Features of a malignant neoplasm include a pleomorphic cell population that retains common morphologic features. Anisocytosis, anisokaryosis, multinucleation with anisokaryosis, nuclear molding, and a high nuclear/cytoplasmic ratio often are observed. Neoplastic cells often have a finely granular chromatin pattern with prominent and sometimes irregular nucleoli. Increased mitotic activity may be present and some mitoses may appear bizarre. In addition, the cells may contain unusual granules, pigment, or express unusual phagocytic activity. The third step is to determine the cell lineage based on characteristic morphologic features.

**Cytology of specific anatomic sites**

The cytology of specific anatomic sites follows. Those tissues and organs that are most frequently sampled during diagnostic work-ups in a clinical setting are discussed. As a result, this article is not exhaustive, but presents an evaluation of the most commonly encountered clinical specimens.

*Skin, subcutis, feather follicles, and emerging feathers*

Normal skin is composed of stratified squamous epithelium that lines the surface of the skin and the feather follicles. Squamous epithelium also is found in emerging feathers but not mature feathers. The dermis is usually thin and consists of collagen fibers and scattered small blood vessels. The subcutis also is very thin in avian skin and consists of lipocytes and fibrocytes.

Abnormal cytologic findings in cutaneous lesions may be diverse. Microscopic changes include various inflammatory cell infiltrates that often have a heterophilic component; phagocytosed bacteria (especially *Staphylococcus* sp); fungal hyphae; yeasts; parasites, such as mites; and various neoplasms. Viral inclusions (circovirus, avian polyomavirus, and poxvirus) may be seen, but are rarely observed cytologically. This section describes more commonly observed changes in the skin and subcutis.

*Knemidokoptes* sp mites are similar in appearance to sarcoptic mange mites of dogs (Fig. 3B). These parasites are associated with hyperkeratosis of unfeathered skin that often involves the legs (scaly leg); feet; and face, including the cere. Routine skin scraping techniques using a scalpel blade and immersion or mineral oil may demonstrate the parasites in heavy mite infestations.

Aspergillosis may involve subcutaneous air sacs, causing swelling and tenderness. Aspirates may contain fungal hyphae that are approximately 4 μm in diameter with thin parallel walls, septa, and right angle branching (Fig. 4A). If the lesion is cavitated, macroconidia may be observed that consist of a round structure that is covered with oval to round fungal spores (Fig. 4B).

The most common neoplasms of the skin and subcutis include lipid-containing masses, spindle cell tumors, lymphoid neoplasms, squamous cell carcinoma, and myelolipoma. Many of these neoplasms can be diagnosed cytologically.

Lipid-containing masses are usually lipomas or xanthomas. Lipomas are benign neoplasms of well-differentiated lipocytes and the smears appear

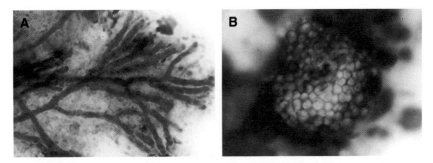

Fig. 4. Appearance of *Aspergillus* sp in cytologic aspirates. (*A*) Fungal hyphae have parallel walls with septa and branching. (*B*) Spore-covered macroconidium from a cavitated lesion in a clavicular air sac; bald eagle (Wright's stain).

greasy or wet when aspirates are smeared on a glass slide. Much of the lipid may be removed by alcohol-based Romanowsky's stains; however, the presence of lipid may be indirectly verified by the application of new methylene blue dye. For clinical use, new methylene blue dye is dissolved in saline. Because fat and water do not mix in the absence of an emulsifying agent, lipid appears refractile and light yellow. The smears consist of free lipid droplets and mature lipocytes distributed singly and in aggregates. Xanthoma is the result of excessive lipid deposits within the skin and subcutis, usually associated with a high-lipid diet, such as excessive sunflower seeds. These masses are firm, spongy, and mobile. Aspirates often produce an oily material that is somewhat gritty. Cytologically, these aspirates consist of free lipid, cholesterol crystals, vacuolated macrophages, and multinucleated giant cells (Fig. 5).

Spindle cell tumors include fibroma, fibrosarcoma, hemangioma, and hemangiosarcoma (Fig. 6). Fibromas have a uniform population of fibrocytes

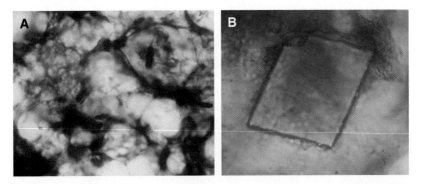

Fig. 5. (*A*) Cluster of lipid-laden macrophages from a xanthoma; budgerigar (Wright's stain). (*B*) High magnification of a cholesterol crystal with typical parallelogram conformation and notch missing in one corner of the crystal (new methylene blue stain).

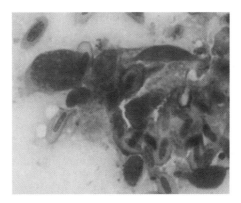

Fig. 6. Pleomorphic spindle cells of fibrosarcoma in an aspirate from a wing mass; cockatiel (Wright's stain).

with an elongated nucleus; mature coarse chromatin pattern; unapparent nucleoli; and attenuated, light blue cytoplasm. In contrast, fibrosarcomas have a more pleomorphic cell population characterized by anisocytosis, anisokaryosis, and scattered mitoses. Individual neoplastic cells have plump, oval nuclei with a finely granular chromatin pattern, multiple nucleoli, and darker blue cytoplasm. Vasoformative neoplasms, such as hemangioma and hemangiosarcoma, consist of blood-filled spaces that are lined by endothelium and supported by minimal collagenous stroma. Both tumors are soft, deformable, and appear blue to purple grossly. Hemangiomas are benign neoplasms in which the blood-filled spaces are lined by well-differentiated, flattened endothelial cells. Aspirates of these neoplasms produce an abundance of blood with thrombocytes aggregates, but endothelial cells are infrequently visualized. Hemangiosarcomas are malignant vasoformative neoplasms consisting of blood-filled spaces lined by pleomorphic angioblasts, frequently in multiple layers, and more solid masses of tumor cells. Aspirates of these neoplasms consist of blood with widely scattered, pleomorphic angioblasts. Individual neoplastic cells have a round to oval to elongate nucleus; moderately granular chromatin; prominent nucleoli; and dark blue cytoplasm with many fine, clear vacuoles. Anisocytosis and anisokaryosis may be prominent. Because these neoplasm have a propensity for hemorrhage, hemosiderin-laden macrophages also may be observed. Aspirates of granulation tissue may closely resemble spindle cell neoplasms. Granulation tissue consists of immature capillaries and proliferation of fibroblasts. Immature fibroblasts have a plump oval nucleus; finely granular chromatin pattern; prominent nucleoli; and attenuated blue, granular cytoplasm. If cytologic doubt exists as to the origin of the mass, a biopsy should be performed.

Depending on the species of bird, lymphoid neoplasms may be composed of a uniform population of lymphoblasts (eg, lymphoid leukosis of chickens), or a mixture of large lymphoblasts, smaller more mature lymphocytes, and plasma cells (eg, Marek's disease of chickens) (Fig. 7) [16]. In companion birds, most lymphomas are of the lymphoblastic type. Although these neoplasms are caused by retrovirus (lymphoid leukosis) and herpesvirus (Marek's disease) infections in chickens, the etiology has not been ascertained in companion birds, although retroviruses may be involved.

Myelolipomas may be found in the skin or liver of companion birds [17]. As the name suggests, these neoplasms are composed of a mixture of lipocytes and hematopoietic cells. Cytologically, they resemble bone marrow aspirates with scattered aggregates of lipocytes and mixed erythroid and myeloid precursors (Fig. 8).

Squamous cell carcinomas usually arise on the unfeathered skin of the face, cere, commissure of the mouth, and uropygial gland. Neoplastic squamous cells exhibit anisocytosis, anisokaryosis, and occasional multinucleation. Individual cells have a large, round nucleus; moderately granular chromatin; and prominent nucleoli. The cytoplasm may appear dark blue with a waxy texture and have perinuclear clearing (halo formation) (Fig. 9).

Other neoplasms, such as renal carcinoma, may metastasize to the skin, but specific diagnosis of such neoplasms is more problematic and usually requires biopsy and histologic examination of the lesion [18]. Cytologic examination can usually classify the neoplasm as of epithelial or mesenchymal origin, however, and suggest whether it is benign or malignant.

Viral inclusions are rarely observed in cutaneous cytology preparations. Poxvirus infection is often associated with small masses involving the eyelids and palpebral conjunctiva. Bollinger's bodies in the cytoplasm of poxvirus-infected cells may appear granular and slightly eosinophilic. Karyomegaly may accompany polyomavirus infection of feather or uropygial gland squamous epithelial cells. Smears of pulp cavity material from emerging feathers may reveal multiple, globular, circoviral inclusions in macrophages, but inclusions may be difficult to distinguish from large keratohyaline granules.

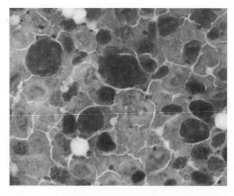

Fig. 7. Large lymphoblasts, necrotic debris, and erythrocytes in hepatic lymphoma; cockatoo (Wright's stain).

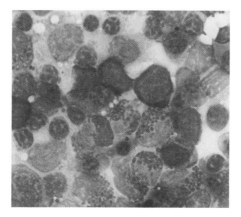

Fig. 8. Aspirate of a myelolipoma containing developing erythroid and myeloid cells; cockatiel (Wright's stain).

## Conjunctiva

The palpebral conjunctiva is lined by squamous epithelium to columnar epithelium, and the bulbar conjunctiva by columnar epithelium. The most common abnormal cytologic findings from lesions in this anatomic site are inflammation, infection, and neoplasia. Inflammatory cell infiltrates have been discussed previously. Common infectious agents include bacteria and *Chlamydophila psittaci*. In chlamydophilosis, scattered macrophages and epithelial cells may contain aggregates of elementary bodies that stain dark purple to gray (Fig. 10). Viral inclusions are rare but may include cytoplasmic Bollinger's bodies of poxvirus infection or marked karyomegaly associated with cytomegalovirus infection. Heterophilic infiltrates may be prominent in bacterial infection, chlamydophilosis, or poxvirus infection.

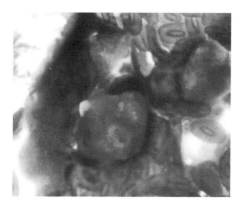

Fig. 9. Squamous cell carcinoma of the uropygial gland. Neoplastic squamous cells are surrounded by blood; cockatiel (Wright's stain).

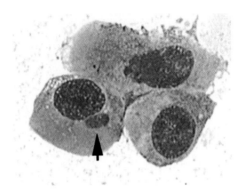

Fig. 10. *Chlamydophila psittaci* inclusion in an epithelial cell. (*arrow*) The elementary bodies are aggregated in the cytoplasm and stain gray-purple; parrot (Wright's stain).

*Respiratory tract*

*Upper respiratory tract, including nasal sinuses, infraorbital sinuses, and trachea*

The sinuses are lined by nonkeratinized squamous cells, ciliated epithelial cells, and goblet cells. The trachea is lined by pseudostratified, ciliated columnar epithelium with scattered goblet cells. A small amount of mucus also may be present. The syrinx is located at the bifurcation of the main stem bronchi and is a common site for restriction or blockage of airflow to the lower airways because of foreign bodies or masses of fungal growth. Abnormal cytologic findings include the presence of inflammatory cell infiltrates, various bacteria including *C psittaci*, *Mycoplasma gallisepticum*, yeasts, fungi, parasites, and foreign bodies. In canaries, poxvirus infection may produce masses in the terminal bronchi and parabronchi. The morphologic appearance of bacteria and *C psittaci* has been covered previously. Mycoplasma infection is especially difficult to diagnose cytologically because the organisms are so small and infected cells have a blue-gray, dusty fringe to the plasma membrane. Yeasts and fungi, such as *Candida* sp and *Aspergillus* sp, have been mentioned previously. *Cryptococcus* sp (including *C uniguttulatus*, *C laurentii*, *C albidus*, *C neoformans* var. *neoformans*, *C neoformans* var. *gattii*, and *C neoformans* var. *grubii*) are often associated with bird droppings and can be cultured from avian cloacal swabs [19]. Generalized avian infection with this dimorphic fungus is rare, however, but has been reported [20]. These organisms vary in size from 5 to 40 μm; may display narrow-based budding; and have a thick, nonstaining, mucopolysaccharides capsule (Fig. 11). If present, these yeasts can be easily discerned in Romanowsky's-stained preparations. Parasites in this anatomic location may include lung mites (especially in finches), trichomonads, and cryptosporidia (Fig. 12). Mites may be observed with subgross magnification. In cytologic preparations, they may be accompanied by pigmented excreta.

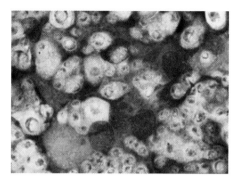

Fig. 11. *Cryptococcus neoformans* is often isolated from avian droppings and cloacal swabs, but rarely causes generalized infection in birds. When observed cytologically, yeasts vary in size from 5 to 40 μm and have a thick, nonstaining capsule (Wright's stain).

*Cryptosporidium baileyi* is more commonly observed in intestinal imprints, but may cause respiratory infection in immunosuppressed birds [21,22]. Small developing to mature sporocysts may be observed within parasitophorus vacuoles in the apical portions of respiratory epithelial cells in tracheal touch imprints obtained at necropsy. The developing sporocysts stain dark blue with Romanowsky's staining, whereas the mature sporocysts appear pink and slightly granular. Foreign bodies usually consist of aspirated portions of seeds.

*Lower respiratory tract, including lung and air sac imprints*

Normal cytologic findings include the presence of ciliated columnar to cuboidal epithelial cells in the terminal bronchioles and air sacs. Goblet cells

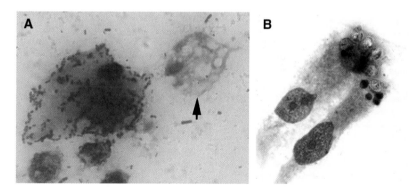

Fig. 12. Protozoa may be challenging to identify in routinely stained cytologic preparations. (*A*) Trichomonad from the oral cavity (*arrow*) appears vacuolated and stains poorly. A squamous cell colonized with bacterial flora also is present; mourning dove (Wright's stain). (*B*) Developing sporocysts of *Cryptosporidium baileyi* in the apical portion of two respiratory epithelial cells. The mature sporocysts stain pink and the immature stages stain dark blue; tracheal imprint, chicken (Wright's stain).

and mucus may be observed in the terminal bronchioles. Nonkeratinized squamous cells line the atria of the lung and portions of air sacs. Abnormal cytologic findings include cellular infiltrates; bacteria (including *C psittaci*); fungus (especially *Aspergillus* sp); and parasites, such as mites, microfilaria, and merozoites of *Sarcocystis falcatula*. In birds with *Mycobacterium avium* complex infection, organisms appear as slightly refractile, unstained rods in the cytoplasm of macrophages or scattered within the background of the imprint. Microfilariae are more common in certain free-ranging birds and have a similar appearance to microfilaria of dogs (Fig. 3A). The meronts of *S falcatula* may be observed as aggregates (meront) within endothelial cells or scattered singly in the background of the imprint. These merozoites are crescent-shaped and have a small nucleus (Fig. 13) [23].

## Gastrointestinal tract

### Oral cavity, choanal slit, esophagus, and crop

These structures are lined with squamous epithelium and possess normal flora. Normal cytologic findings include keratinized and nonkeratinized squamous cells, bacteria, and ingested plant and seed material. The normal bacterial flora may include *Alysiella filiformis*. This bacterium is often attached to squamous cells in specimens obtained from the upper alimentary tract and appears as ribbons or chains of coccobacilli. This organism is related to *Simonsiella* sp, which is an oral inhabitant of cats and dogs [24,25]. Abnormal cytologic findings include inflammatory cell infiltrates, yeasts, and parasites or parasitic ova. The character of the inflammatory infiltrate varies with the disease process. *Candida* sp are oval and stain blue with Romanowsky's stains. These organisms may also be accompanied by budding; germinal tube extension (pseudohypha formation); or the production of hyphae (Fig. 14). *Trichomonad* sp is usually associated with oral masses, especially in doves (see Fig. 13). The protozoa have a small nucleus and flagella

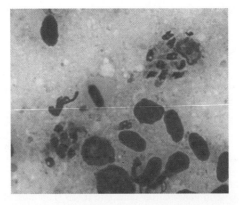

Fig. 13. Merozoites of *Sarcocystosis falcatula* in a lung imprint; cockatoo (Wright's stain).

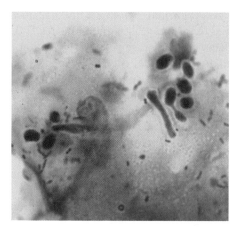

Fig. 14. *Candida* sp may appear as oval yeasts or exhibit pseudohypha (and true hypha) formation; Amazon parrot (new methylene blue stain).

that may be difficult to visualize. In many instances, the cytoplasm of these organisms is highly vacuolated. The ova of *Capillaria* sp are barrel-shaped and possess bipolar plugs (Fig. 15). These ova may be differentiated from pollen grains that appear round with a rough exterior and a single point of attachment.

## Cloaca and vent

The cloaca is lined with columnar to transitional epithelial cells. Variable numbers of goblet (mucous) cells also may be present. The vent is lined by stratified squamous epithelium. Normal cytologic findings include plant and food debris, mixed bacteria, and mucus. Urates are commonly observed in

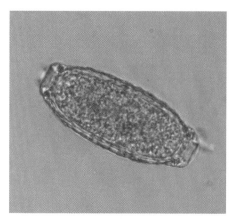

Fig. 15. Ova of *Capillaria* sp appear barrel-shaped with a plug at each end (unstained wet mount, zinc sulfate flotation).

cloacal preparations. Abnormal cytologic findings include inflammatory cells, certain bacteria, fungi and yeasts, and parasites. The type of inflammatory cells is dependent on the disease process. Detection of offending bacteria is more problematic. Clostridial enteritis is associated with the presence of large bacilli with clear, round spores that may assume a safety pin appearance (Fig. 16). These organisms can be discerned on Romanowsky's-stained smears and appear dark blue in Gram-stained smears. Gram stain also may be used to visualize excessive gram-negative bacteria that appear as red rods. Yeasts of *Candida* sp are described previously. Another pathogen in cloacal swabs is the fungus *Macrorhabdus ornithogaster* [26]. This organism infects the proventricular-ventricular junction but scattered organisms may be observed in cloacal swabs. This pathogen was previously known as "megabacterium." It is approximately 1 mm in length and stains faintly blue with Romanowsky's stain, dark blue with Gram stain, and bright pink by the periodic acid–Schiff technique (Fig. 17). This organism is elongate and sometimes gently bowed with parallel sides and blunt ends. *M ornithogaster* must be distinguished from feather barbules. Feather barbules are reminiscent of sections of a car antenna. They are approximately the same length as *M ornithogaster* but are straighter, thinner, and have cup-shaped termini.

Parasites may be serendipitous findings in routine cytologic preparations. These include various parasite ova, small sporocysts, sporozoites, and flagellates.

*Body cavity effusions*

Birds have a single body cavity that is partially compartmentalized by transparent membranes. Smears of body cavity fluid usually consist of scattered macrophages and a few mesothelial cells, but the protein background is seldom noticeable. Ciliated cuboidal cells from air sacs may be visualized

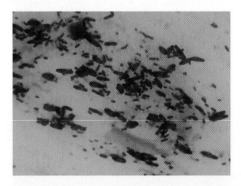

Fig. 16. *Clostridium* sp in a cloacal swab specimen are large, oval-to-elongate bacteria with a lighter area of spore formation that gives the organisms a safety pin appearance; clostridial enteritis, lorikeet (Gram stain).

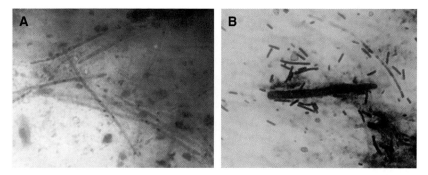

Fig. 17. *Macrorhabdis ornithogaster*, previously known as "megabacterium," are gastrointestinal fungi that infect the proventricular-ventricular junction. (*A*) Organisms stain pale blue with Wright's stain. (*B*) The same fungi appear dark blue following the application of Gram stain; canary (Wright's and Gram stains).

infrequently. Abnormal cytologic findings include cellular infiltrates with bacteria or fungi, egg yolk, or neoplastic cells. Egg yolk coelomitis is more common in older hens. Body cavity fluid often appears yellow to cream-colored with a flocculent texture. Microscopically, the smears contain macrophages, heterophils, mesothelial cells, and infrequent multinucleate giant cells embedded within a homogeneous to granular to globular, pink to blue, amorphous background material (yolk) (Fig. 18). Carcinoma cells may be observed in body cavity effusions on rare occasions. Most of these specimens are from hens with ovarian, oviductal, or intestinal carcinoma. Individual cells may appear round, cuboidal, or columnar. The nucleus is round and may have a visible nucleolus. The cytoplasm is scant to moderately abundant and may range from blue-gray to pink.

*Synovial fluid*

Synovial fluid is difficult to obtain from most birds and only a small drop of material usually is available for cytologic examination. Normal cytologic

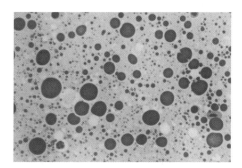

Fig. 18. Free egg yolk may appear as eosinophilic to basophilic globules with minimal inflammatory cell infiltration. Coelomic cavity; Amazon parrot (Wright's stain).

findings include widely scattered mononuclear cells in a pink protein precipitate. Abnormal findings include excessive numbers of mononuclear cells or heterophils and various bacteria (especially *Staphylococcus* sp). Gout may occasionally be diagnosed cytologically in birds. Aspirates from affected joints of birds with gout may yield a small amount of chalky white material or uric acid crystals. Under polarized light, this material may be refractile and have needle-like configurations (Fig. 19). Because uric acid is water soluble, this material seldom survives routine Romanowsky's staining unless an inordinately large quantity of uric acid crystals is present.

## Liver and spleen

Aspiration and endoscopic biopsy of the liver and spleen are becoming more common in diagnostic avian medicine. Touch imprints of small biopsies or of either organ may provide information about the disease process in a given individual. Common diseases of these organs that can be diagnosed cytologically are discussed.

## Liver

The normal liver consists of cords of hepatocytes separated by blood-filled sinusoids that contain scattered macrophages (Kupffer cells). The blood supply consists of arterial and portal blood inflow and hepatic vein outflow. Bile produced by hepatocytes is excreted into canaliculi that communicate with the bile duct and exit into the duodenum. Depending on the species of bird, a gallbladder may or may not be present. Normal aspirates or touch imprints contain blood, clusters of hepatocytes, a few macrophages, and occasional aggregates of biliary epithelial cells. Hepatocytes are polygonal with a round nucleus; single nucleolus; and gray, granular cytoplasm. Biliary epithelial cells are cuboidal and have a round, hyperchromatic nucleus and light blue cytoplasm. Kupffer cells may occasionally contain particles of hemosiderin. Hepatic lipidosis is a common finding in

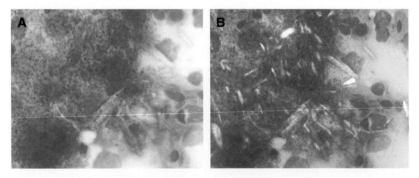

Fig. 19. Appearance of uric acid crystals in an aspirate from the hock joint of a chicken. (*A*) Remaining uric acid crystals are difficult to discern in Wright's stained preparation. (*B*) Scattered needle-shaped crystals are readily visualized with polarized light.

cytologic specimens from liver. Affected hepatocytes contain clear, round, cytoplasmic vacuoles with Romanowsky's staining. In severe hepatic lipidosis, hepatocytes may be very difficult to identify (Fig. 20). Much of the lipid may be removed during Romanowsky's staining because the dyes are dissolved in absolute methanol. If necessary, lipid may be demonstrated indirectly with new methylene blue stain (a water-based stain) or directly with oil red O or Sudan (III or IV) staining. Hepatic lipidosis of very young birds is associated with normal yolk metabolism. In older birds, hepatic lipidosis may be a reflection of a lipid-rich diet or associated with metabolic disease. Abnormal cytologic findings include Kupffer cells with excessive erythrophagocytosis or hemosiderin. Hemosiderin may appear golden brown to blue in Romanowsky's-stained preparations. Increased iron accumulation within Kupffer cells or increased erythrophagocytosis may be observed in iron overloading or with infection where accelerated destruction of erythrocytes occurs. Excessive iron storage within hepatocytes may give the cytoplasm a grayish to brown, granular appearance. In many cases of iron accumulation, the hemosiderin may appear as nondescript gray granules (Fig. 21). Diagnostic pathologists can use Perls' technique, however, to demonstrate to the presence of iron. Hemosiderin particles subsequently stain bright blue (see Fig. 21). Hepatocellular necrosis may be difficult to identify cytologically. The cells may stain blue-gray and lack nuclear detail. Inflammatory cell infiltrates may be present following blunt trauma, bacterial infection, chlamydophilosis, viral infection, hepatocellular necrosis, or parasitism. In mycobacteriosis, macrophages are frequently numerous and organisms may be present within macrophages or scattered throughout the imprint. Parasites that are most commonly encountered in cytology preparations include shizonts of *Plasmodium* sp and lymphocytes containing gametocytes of *Atoxoplasma* sp. Plasmodial shizonts are large and round with a pink to purple, spotty appearance (Fig. 22). Gametocytes of *Atoxoplasma* sp infect lymphocytes [27]. These organisms are located in the

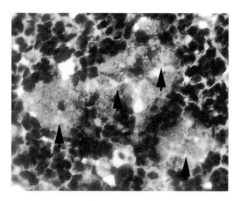

Fig. 20. (*arrows*) Four lipid-laden hepatocytes and blood in an aspirate from a bird with severe, dietary-associated hepatic lipidosis; Amazon parrot (Wright's stain).

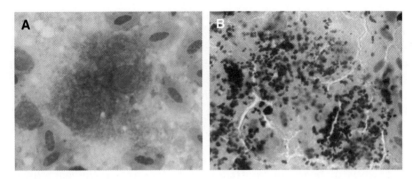

Fig. 21. Hemosiderin accumulation within hepatocytes. (*A*) Iron particles appear gray-brown and are difficult to visualize in the Wright's-stained liver imprint. (*B*) Application of Perls' stain reveals innumerable, bright blue iron granules within affected hepatocytes. Hepatic hemosiderosis; Bali mynah (Wright's and Perls' stains).

cytoplasm of the cell but often partially displace the cell's nucleus reminiscent of a cookie with one bite removed (Fig. 23). Foci of nematode migration may be rich in macrophages and multinucleated giant cells. Although viruses often target hepatocytes, viral inclusions are rarely visible in routinely stained preparations. Exceptions are some aviadenovirus infections wherein marked karyomegaly is apparent (Fig. 24). The nuclear inclusions have a stippled or marbled appearance cytologically. In avian polyomavirus infection, karyomegalic nuclear inclusions are blue-gray and glassy. In chronic hepatic disease, extramedullary hematopoiesis may occasionally be observed, especially with regard to the heterophil lineage. Excessive bile duct epithelial cells may be observed with biliary hyperplasia or bile duct carcinoma. Desmoplasia often accompanies bile duct carcinoma, so immature fibroblasts also may be observed. Increased numbers of lymphoblasts should alert the microscopist to the possibility of lymphoma.

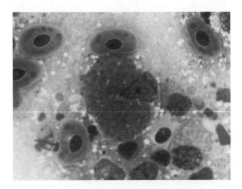

Fig. 22. *Plasmodium relictum* shizont in a liver imprint accompanied by piroplasms within erythrocytes; African black-footed penguin (Wright's stain).

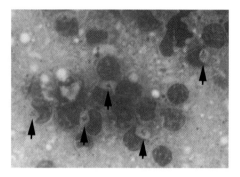

Fig. 23. *Atoxoplasma* sp gametocytes (*arrows*) in the cytoplasm of lymphocytes in an hepatic imprint; Bali mynah (Wright's stain).

Secondary (reactive or AA-type) amyloidosis may occur as a sequel to chronic inflammation or infection [28]. Amyloid usually is deposited in the liver, spleen, or joints. This condition is more common in birds from the orders Anseriformes (especially Pekin ducks), Gruiformes, and Phoenicopteriformes, but may occur in any avian species. Although amyloidosis may occur in birds, especially in ducks, this substance may be difficult to recognize cytologically. Amyloid usually appears as a bright pink, amorphous, extracellular material.

*Spleen*

Histologically, the spleen is bordered by a thin capsule of fibrous connective tissue containing infrequent smooth muscle cells [29]. The parenchyma consists of blood-rich red pulp and lymphoid cell-rich white pulp. Normal cytologic preparations have an abundance of blood admixed with a heterogeneous population of lymphocytes and plasma cells. Cytologic abnormalities include excessive erythrophagocytosis or hemosiderin concentration within macrophages; the presence of inflammatory infiltrates;

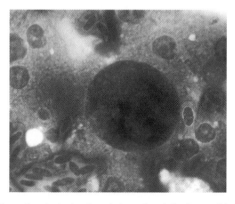

Fig. 24. Karyomegalic nuclear inclusion in aviadenovirus infection; red-bellied parrot (Wright's stain).

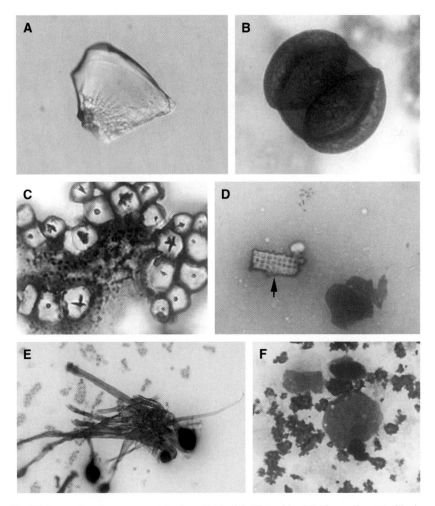

Fig. 25. Examples of common cytologic artifacts. (*A*) Glass chip. (*B*) Pine pollen. (*C*) Gloving powder granules. (*D*) Gauze square fragment (*arrow*). (*E*) Plant fibers. (*F*) Ultrasound paste (*flocculent purple material*).

extramedullary hematopoiesis; bacteria including mycobacteria and *C* (formerly *Chlamydia*) *psittaci*; parasites, such as *Atoxoplasma serenii*; the presence of numerous lymphoblasts (lymphoma); and amyloid deposition.

## Common artifacts and problem areas in avian cytology

### Common artifacts

Artifacts in avian cytology preparations are numerous. They may be mistaken for pathogens or obscure significant findings and should be avoided.

A short list of such artifacts follows (Fig. 25). Glass chips from microscope slides and coverslips may resemble crystalline material, such as oxalate crystals. Pollen grains may be confused with some parasite ova. Erythrocyte refractile artifact, caused by water or air between the cell surface and immersion oil layer, may be mistaken for parasites. Some forms of stain precipitate may mimic or obscure bacteria. Talc or starch granules may be confused with in vivo crystal formation; they appear hexagonal, colorless, and refractile with a central depression or navel. One of the greatest detriments to cytologic examination is ultrasound paste or surgical lubricant. These materials appear deep pink to purple and flocculent to granular. They cover cellular material and greatly interfere with microscopic examination of cytologic specimens.

*Potential problem areas in cytology*

There are a few situations in which a microscopist may make grave errors in diagnostic cytology. The foremost is misdiagnosis of broken cells as neoplasia. An intact cell should have clearly defined nuclear and plasma membranes. If the cell is lysed the nuclear DNA spreads and the nucleoli may be visually prominent. Lymphoma may be difficult to distinguish from marked lymphoid hyperplasia. In such lesions, a biopsy may be warranted. In coelomic cavity preparations, mesothelial cell hyperplasia may be difficult or impossible to distinguish from carcinomatosis or mesothelioma. Finally, reactive fibroplasia may be misdiagnosed as fibrosarcoma. This is especially true in avian lesions where reactive fibroplasia may appear bizarre. For veterinary clinicians, it is also important to know one's cytologic limitations and seek the opinion of a board-certified clinical pathologist when necessary.

## References

[1] Campbell TW. Avian hematology and cytology. 2nd edition. Ames (IA): Iowa State University Press; 1995.
[2] Campbell TW. Cytology. In: Ritchie BW, Harrison GJ, Harrison LR, editors. Avian medicine: principles and application. Lake Worth: Wingers Publishing; 1994. p. 199–221.
[3] Rakich PM. Cytology. In: Duncan & Prasse's veterinary laboratory medicine: clinical pathology. 4th edition. Ames (IA): Iowa State Press; 2003. p. 304–30.
[4] Evans EW, Harmon BG. A review of antimicrobial peptides: defensins and related cationic peptides. Vet Clin Pathol 1995;24:109–16.
[5] Brockus CW, Jackwood MW, Harmon BG. Characterization of beta-defensin prepropeptide mRNA from chicken and turkey bone marrow. Anim Genet 1998;29:283–9.
[6] Harmon BG. Avian heterophils in inflammation and disease. Poult Sci 1998;77:972–7.
[7] Latimer KS, Tang K-N, Goodwin ME, et al. Leukocyte changes associated with acute inflammation in chickens. Avian Dis 1988;32:760–72.
[8] Maxwell MH, Burns RB. Blood eosinophilia in adult bantams naturally infected with *Trichostrongylus tenuis*. Res Vet Sci 1985;39:122–3.
[9] Maxwell MH, Burns RB. Experimental eosinophilia in domestic fowls and ducks following horse serum stimulation. Vet Res Commun 1982;5:369–76.

[10] Montali RJ. Comparative pathology of inflammation in the higher vertebrates (reptiles, birds and mammals). J Comp Pathol 1988;99:1–26.
[11] Golemboski KA, Whelan J, Shaw S, et al. Avian inflammatory macrophage function: shifts in arachidonic acid metabolism, respiratory burst, and cell-surface phenotype during the response to sephadex. J Leukoc Biol 1990;48:495–501.
[12] Olah I, McCorkle F, Glick B. Lectin-induced giant cell formation in the chicken wattle. Poult Sci 1980;59:2151–7.
[13] Latimer KS, Bienzle D. Determination and interpretation of the avian leukogram. In: Feldman BF, Zinkl JG, Jain NC, editors. Schalm's veterinary hematology. 5th edition. Baltimore: Lippincott Williams & Wilkins; 2000. p. 417–32.
[14] Doolen MD, Greve JH. Description of a microfilaria from an umbrella cockatoo (*Cacatua alba*) and an unsuccessful attempt to infect mosquitoes (*Culex pipiens pipiens*). Avian Dis 1992;36:484–7.
[15] Latimer KS. Oncology. In: Ritchie BW, Harrison GJ, Harrison LR, editors. Avian medicine: principles and application. Lake Worth: Wingers Publishing; 1994. p. 640–72.
[16] Gregory CR, Latimer KS, Mahaffey EA, et al. Lymphoma and leukemic blood picture in an emu (*Dromaius novae hollandiae*). Vet Clin Pathol 1996;25:136–9.
[17] Latimer KS, Rakich PM. Subcutaneous and hepatic myelolipomas in four exotic birds. Vet Pathol 1995;32:84–7.
[18] Latimer KS, Ritchie BW, Campagnoli RP, et al. Metastatic renal carcinoma in an African grey parrot (*Psittacus erithacus erithacus*). J Vet Diagn Invest 1996;8:261–4.
[19] Rosario I, Hermoso de Mendoza M, Deniz S, et al. Isolation of *Cryptococcus* species including *C. neoformans* from cloaca of pigeons. Mycoses 2005;48:421–4.
[20] Raso TF, Werther K, Miranda ET, et al. Cryptococcosis outbreak in psittacine birds in Brazil. Med Mycol 2004;42:355–62.
[21] Latimer KS, Goodwin MA, Davis MK. Rapid cytologic diagnosis of respiratory cryptosporidiosis in chickens. Avian Dis 1988;32:826–30.
[22] Latimer KS, Steffens WL, Rakich PM, et al. Cryptosporidiosis in four cockatoos with psittacine beak and feather disease. J Am Vet Med Assoc 1992;200:707–10.
[23] McCormick-Rantze ML, Latimer KS, Wilson GH. Sarcocystosis in psittacine birds. University of Georgia College of Veterinary Medicine. Available at: http://www.vet.uga.edu/vpp/clerk/rantze/index.htm. 2003. Accessed November 30, 2006.
[24] Kaiser GE, Starzyk MJ. Ultrastructure and cell division of an oral bacterium resembling Alysiella filiformis. Can J Microbiol 1973;19:325–7.
[25] Xie CH, Yokota A. Phylogenetic analysis of *Alysiella* and related genera of Neisseriaceae: proposal of *Alysiella crassa* comb. nov., *Conchiformibium steedae* gen. nov., comb. nov., *Conchiformibium kuhniae* sp. nov. and *Bergeriella denitrificans* gen. nov., comb. nov. J Gen Appl Microbiol 2005;51:1–10.
[26] Son TT, Wilson GH, Latimer KS. Clinical and pathological features of megabacteriosis (*Macrorhabdus ornithogaster*) in birds. University of Georgia College of Veterinary Medicine. Available at: http://www.vet.uga.edu/vpp/CLERK/Son/index.htm. 2004. Accessed November 30, 2006.
[27] Sheridan KL, Latimer KS. An overview of atoxoplasmosis in birds. University of Georgia College of Veterinary Medicine. Available at: http://www.vet.uga.edu/vpp/CLERK/Sheridan/index.htm. 2002. Accessed November 30, 2006.
[28] Landman WJM, Gruys E, Gielkens ALJ. Avian amyloidosis. Avian Pathol 1998;27:437–49.
[29] Powers LV. The avian spleen: anatomy, physiology, and diagnostics. Comp Cont Educ for Pract Vet 2000;22:838–43.

VETERINARY
CLINICS
Exotic Animal Practice

# Cytologic Diagnosis of Diseases in Reptiles

A. Rick Alleman, DVM, PhD, DABVP, DACVP[a],*,
Emily K. Kupprion, VMD[b]

[a]*College of Veterinary Medicine, University of Florida, 2015 SW 16th Avenue, Room V2-240, Gainesville, FL 32610, USA*
[b]*Animal Emergency Hospital-Volusia, 3 Bimini Circle, Ormond Beach, FL 32176, USA*

The class Reptilia encompasses thousands of species grouped as chelonians, crocodilians, squamates, and the Rhynchocephalia. This group of vertebrates is extremely diverse in terms of the anatomy and physiology of its members. The diversity of the species within this class can make the job of the reptile practitioner challenging. Cytology can be a valuable tool to aid the practitioner in identifying specific disease processes in masses, affected organ tissues, and body fluids. The value of cytology in the rapid identification of various disease processes may be obvious. The principles used in the application of cytology can be broadly applied to various species of mammals and reptiles. The accurate interpretation of cytologic specimens requires microscopic skills and the ability to understand and recognize reptilian cell types involved in various disease processes, particularly neoplasia and inflammation.

This article provides basic information necessary for the collection, preparation, and evaluation of cytology specimens collected from tissue lesions, organs, and body fluids in reptiles. The critical rule for a practicing cytologist is not to overinterpret the cytologic findings. For example, if you are sure that the lesion is a neoplasm, based on the cellular composition of a lesion, your cytologic interpretation would be a "neoplastic process." If some of the nuclear features are typical of a malignant tumor, your interpretation would be a "malignant neoplasm." If the cytoplasmic features and arrangement of the cells in the preparation indicate a malignant tumor of mesenchymal origin, your diagnosis would be "sarcoma." Finally, if the location of the lesion and characteristic features of the cytologic preparation fit with

* Corresponding author.
*E-mail address:* allemanr@mail.vetmed.ufl.edu (A.R. Alleman).

a malignant mesenchymal tumor of osteoblast origin, your diagnosis could be "osteosarcoma." Notice that if the lesion were, in fact, an osteosarcoma, any of these interpretations would be correct. As long as the interpretation is accurate, cytology serves as an important part of the diagnostic workup. The preciseness of the answer is not as important as the accuracy. Each interpretation is different, and the ability to define exactly what a lesion is, or what caused it, varies depending on the sample collected, the quality of the preparation, and the skills of the interpreter. Nevertheless, any information provided is useful as long as it is accurate.

## Sample collection and slide preparation

Fine-needle aspirates (FNAs) are used to obtain cells from solid tissue masses, organs, or body cavities. This technique is particularly useful in sampling skin lesions but is also commonly used for intracavitary lesions and fluids. After preparation of the skin surface with alcohol, a 22- to 25-gauge 1- to 1.5-in needle attached to a 6-mL syringe is inserted into the lesion. Once the needle tip is in the lesion or organ, the plunger of the syringe is drawn back, producing a vacuum. If aspirating a tissue mass, manipulation of the needle position and rapid repositioning of the needle tip by gentle movements within the lesion while maintaining vacuum negative pressure on the syringe are ideal. This allows sampling from different areas of the lesion in case the mass consists of a heterogeneous population of cells. Five to six redirections of the needle are generally sufficient. The plunger and associated negative pressure are released just before withdrawing the needle from the tissue. To prepare the sample, smears are made on clean glass slides. Glass slides should be readily available once the sample is collected. Aspirated material tends to clot quickly, and the presence of platelet and fibrin clumps adversely affects smear preparation and cytologic evaluation. Once material appears in the hub of the needle, the procedure is discontinued and the needle is removed from the lesion. The syringe is then detached from the needle, and air is drawn into barrel of the syringe. The syringe is reattached to the needle, and gentle pressure is applied to the plunger in an effort to force the cells from within the needle onto the glass slide. Once a droplet of material is visualized on the tip of the needle, the droplet is placed onto one end of the glass slide and a second slide is positioned on top of the sample slide (Fig. 1). The material is gently sandwiched between the two slides, allowing the material to be spread onto a monolayer of cells by the weight of the top slide (Fig. 2). Gentle traction in opposite directions is applied, and the two slides are pulled apart in a horizontal plane, producing two cytologic preparations that are ready for staining and cytologic evaluation.

Skin scrapings are commonly performed to evaluate cutaneous lesions in reptile species. They are especially useful for lesions with eroded surfaces. A dulled sterile scalpel blade or spatula may be used to scrape the surface of

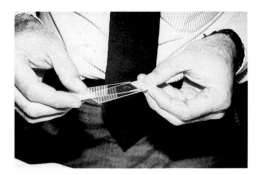

Fig. 1. A droplet of aspirated material is placed on one end of a clean glass slide.

the lesion gently. Depending on the consistency of the lesion, few or many passes over the tissue may be needed to obtain an adequate sample. The sample is deposited onto a clean glass slide by scraping the blade edge onto the slide. The collected material should be gently swiped once over the length of the slide, allowing an even smear of variable thickness. If excess material is available, similar spreads should be made on different slides. The slides are then air dried and stained. Alternatively, a drop of mineral oil may be applied to the scraped material to spread the sample around on the slide. After mixing the mineral oil and the material together, the slide may be coverslipped for immediate examination. This technique does not require staining and is particularly useful for identification of parasites, such as mite or nematode infestations, but does not allow enough cytologic detail to evaluate other disease processes, such as inflammation or neoplasia.

Impression smears of samples collected by an incisional or excisional biopsy can also be prepared from tissues before placing the sample into formalin. After the sample is obtained, the cut surface of the lesion is blotted once on a clean towel or napkin and then gently imprinted several times

Fig. 2. A second slide is quickly placed on top of the sample slide to allow the material to spread evenly between the two slides. The two slides are then gently slid apart in a horizontal plane.

in different locations on the microscope slide (Fig. 3). The slide is then ready for staining and cytologic evaluation.

Swabbed preparations may be made from material collected from oral lesions, fecal material, cloacal contents, tracheal material, draining tracts, conjunctiva, or nasal discharge. Swabs of various sizes may be used depending on area to be sampled and size of the patient. The swab is gently touched and rolled against or into the affected area to obtain exfoliated cells, pathogens, or discharge material. After the sample is obtained, the swab is rolled along a glass slide. The sample is air dried and then stained for cytologic evaluation.

Once a slide is air dried, the material is adherent to the slide surface and the slide may be stained. Cytologic preparations are typically stained with a Romanowsky-type stain; however, certain selective situations warrant evaluation with a new methylene blue stain. One Romanowsky-type stain commonly used in clinical practice is Diff-Quik (Diff-Quik Stain Set; Dade Behring, Deerfield, Illinois). Similar three-step staining solutions are available, however. In general, Romanowsky-type stains are affordable, easy to obtain, require minimal maintenance, and are not difficult to use. They stain cell cytoplasm and microorganisms well and provide adequate nuclear and nucleolar detail sufficient for most routine cytologic evaluations. Diff-Quik (or other commercially available rapid use stains) has its own unique recommended staining procedures [1]. In general, all samples should be placed in the aqua-colored fixative solution for a minimum of 1 minute [2]. Proper fixation is essential for quality staining. After fixation, the slide is slowly dipped in and out of the eosinophilic stain for 8 to 10 dips. The excess stain is absorbed onto a paper towel, and the dipping procedure is repeated in the basophilic stain. This procedure should be adapted to the thickness of the preparation and type of sample. For example, a thick preparation of a highly cellular tissue aspirate requires a longer staining period to achieve adequate penetration of the stain than would a smear of a poorly cellular aspirate or fluid preparation. Conversely, a sample that

Fig. 3. The cut surface of a tissue lesion is lightly blotted on an absorbent surface to remove excess blood. The lesion is then imprinted in several spots on the surface of a clean glass slide.

has been stained too long or has absorbed so much stain that it is too dark to evaluate properly may be placed in methanol for destaining. Once destained, a slide may be "restained" as needed. Commercial and university diagnostic laboratories frequently alter the staining procedure (time) based on the type and thickness of the sample.

New methylene blue, although not routinely used by most cytologists, is useful in selected circumstances. This basic stain does not stain cytoplasm well but provides excellent nuclear and nucleolar detail [1]. In addition, it stains most organisms and tends to highlight fungal and yeast organisms that can easily be missed using Romanowsky-type stains. Because the stain does not require alcohol fixation, lipids and adipocytes are preserved in the staining procedure [2]. Once material is smeared and air dried on a slide, a drop of new methylene blue is placed on top of the material and a dust-free coverslip is overlaid. The preparation is now ready for cytologic evaluation.

## Tissue aspiration: classifying the disease process

The same cytologic techniques used to evaluate samples collected from mammalian species can be broadly applied to the cytologic evaluation of reptiles. The disease processes described here are pathologic processes that affect all animals. With minor modifications to the species-specific cell types found in reptiles, the same types of samples and disease processes encountered in mammalian species are observed in reptile species. The foundation of the cytologic interpretation is, first and foremost, to classify the lesion into one of five categories of basic pathologic disease processes. The five general categories of lesions are as follows: (1) inflammatory lesion, (2) cystic lesion, (3) hemorrhagic lesion, (4) neoplastic lesion, and (5) mixed cell population. Most pathologic processes that cause tissue swelling or damage can be classified into one of these five broad categories. Therefore, if the predominant cytologic feature of each of these categories can be recognized, the lesion can be classified with confidence. Once classified, the practicing cytologist should try to proceed to a more specific diagnosis if more specific cytologic features are recognized. The general cytologic features of each of these categories are described in Table 1.

The major cytologic feature of an inflammatory lesion is the presence of heterophils in greater number than what would be expected from blood contamination. Cystic lesions contain large numbers of mature keratinized squamous epithelial cells or abundant acellular material. Hemorrhagic lesions contain blood and variable numbers of macrophages and phagocytized erythrocytes (erythrophagia) or dark-blue hemosiderin (iron pigment). Neoplastic lesions are characterized by a homogeneous population of cells derived from the same tissue origin. A mixed cell population is a cytologic preparation containing inflammatory cells (heterophils,

Table 1
Categories of lesions from tissue aspirates

| Category | Cytologic features |
| --- | --- |
| Inflammation | Heterophils ± macrophages greater than what is expected from blood contamination ± variable lymphocytes and eosinophils |
| Cystic | Poorly cellular proteinaceous fluid or mature squamous epithelial cells |
| Hemorrhagic | Blood and macrophages exhibiting erythrophagia or containing hemosiderin or hematoidin |
| Neoplastic | Homogeneous cell population derived from the same tissue origin; further classified into round cell, epithelial, mesenchymal, or neuroendocrine |
| Mixed cell population | Inflammation + noninflammatory population (epithelial or mesenchymal cells) |

macrophages, lymphocytes, eosinophils, or basophils) and noninflammatory cells (epithelial or mesenchymal cells). These five general categories of lesions are described in detail in the next sections.

In cases in which more than one pathologic process is identified, the lesion is classified according to the primary or inciting cause. For instance, there may be hemorrhage within a neoplasm or inflammation within a cyst.

## Inflammatory lesions

Two types of inflammation commonly occur, purulent inflammation and pyogranulomatous inflammation. They are distinguished from each other by the type of inflammatory cells present or the absence of other specific cell types. The type of inflammatory response often gives an indication as to the disease process that caused it.

Purulent inflammation is characterized by the presence of a predominant (>85%) population of heterophils (Fig. 4). In any inflammatory response but particularly in purulent inflammation, heterophils should be evaluated

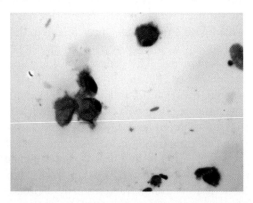

Fig. 4. Tissue aspirate contains four nondegenerate heterophils in a purulent inflammatory response from a water turtle (Wright-Giemsa, original magnification ×50).

for the presence of degenerative and toxic changes. Degenerative changes, such as karyolysis (nuclear lysis) and karyorrhexis (nuclear fragmentation), and toxic changes, such as degranulation and cytoplasmic basophilia, may be an indication of bacterial sepsis (Fig. 5). When observed, careful evaluation of the preparation for microorganisms is indicated. If no organisms are identified, culture for pathogen identification is indicated.

Pyogranulomatous inflammation (or mixed inflammation) is characterized by the presence of heterophils with a 15% or greater percentage of macrophages and, sometimes, lymphocytes (Fig. 6). This is the most common type of inflammation observed in reptile species, because acute heterophilic inflammation rapidly progresses to have a moderate macrophage component [3]. Increased numbers of lymphocytes and plasma cells may also be observed [3]. Inflammation consisting predominantly of macrophages, although typically associated with chronic inflammation, is not pathognomonic for a chronic process. When it is observed in a lesion, however, close examination of a tissue aspirate for identification of a foreign body or fungal hyphae is indicated (Fig. 7). If none are identified and a fungal infection is still suspected, an air-dried slide preparation may be sent to a commercial laboratory for special staining (Gomori's methamine silver stain) that can accentuate any hyphal structures present (Fig. 8).

## Cystic lesions

Cystic lesions in reptiles are frequently associated with parasitism and, less commonly, severe trauma or burns [4]. Sebaceous cysts, cysts of sebaceous gland origin, have also been reported in the lizards, particularly the

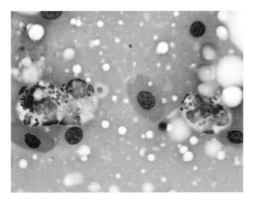

Fig. 5. Tissue aspirate from a green iguana contains three degenerate and toxic heterophils. Phagocytized intracellular bacteria indicate sepsis. The nuclei of all three cells are karyolytic, indicating nuclear degeneration and partial (*cell on right*) or complete (*two cells on left*) degranulation. Degranulation and the cytoplasmic basophilia indicate cytoplasmic toxicity. Four intact erythrocytes are also present (Wright-Giemsa, original magnification ×100). (Courtesy of H.L. Wamsley, DVM, DACVP, Gainesville, FL.)

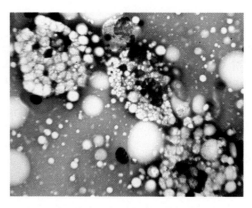

Fig. 6. Tissue aspirate from a green iguana contains three large reactive heavily vacuolated macrophages and one smaller mildly reactive macrophage (*top center*) (Wright-Giemsa, original magnification ×100.) (Courtesy of H.L. Wamsley, DVM, DACVP, Gainesville, FL.)

green iguana [5]. These lesions appear grossly as dry abscesses and cytologically contain an amorphous basophilic material. Cystic fluid from subcutaneous or intracoelomic locations appears cytologically as a proteinaceous poorly cellular fluid, with small numbers of glandular epithelium.

## Hemorrhagic lesions

Hemorrhagic lesions include hematomas from trauma or a bleeding disorder, seromas secondary to trauma or a postoperative procedure, and bleeding tumors (hemangioma or hemangiosarcoma). True hemorrhage must be distinguished from blood contamination attributable to capillary bleeding, which is often associated with sample collection. Cytologically,

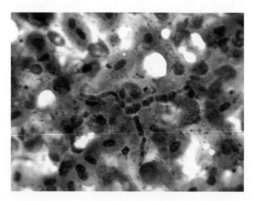

Fig. 7. Tissue aspirate from a rat snake with a fungal granuloma. A septate branching fungal hypha is seen in the center, surrounded by a sheet of epithelioid macrophages (Wright-Giemsa, original magnification ×50).

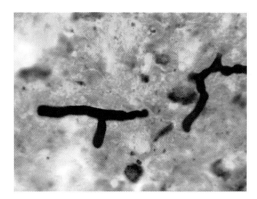

Fig. 8. Tissue aspirate from a rat snake with a fungal granuloma. A septate branching fungal hypha is accentuated using a fungal stain. (Gomori's methamine silver, original magnification ×50).

true hemorrhage is recognized by the presence of abundant erythrocytes and variable numbers of macrophages in which phagocytized erythrocytes or hemosiderin is observed (Fig. 9).

## Neoplastic lesions

Neoplasia is grossly characterized by the presence of abnormal tissue swelling or architecture and is cytologically defined by a homogeneous population of cells. This is best appreciated by the presence of cells with the same nuclear and cytoplasmic staining characteristics. Two important determinations should be attempted if a neoplasm is diagnosed: determine if the lesion is benign or malignant, and determine the tissue of origin. The

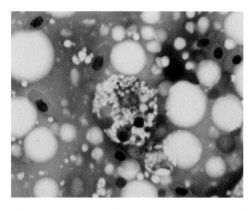

Fig. 9. Tissue aspirate from an iguana contains a large macrophage (*center*) with a phagocytized erythrocyte, indicating hemorrhage in the lesion. A small degenerate heterophil is seen below the macrophages, and small numbers of erythrocytes are noted in the background (Wright-Giemsa, original magnification ×50).

distinction between benign and malignant neoplasms is made by evaluation of nuclear features. The determination of the tissue of origin is accomplished by evaluation of the cytoplasmic features and orientation of the cell population.

The characteristics of benign neoplasia or hyperplasia include uniform cytoplasmic and nuclear size and shape; a uniform nuclear-to-cytoplasmic ratio (N/C ratio); and minimal variation in the size, shape, and number of nucleoli. Tissue aspirates taken from malignant lesions often are more cellular than would normally be expected from the tissue or organ sampled, or they may even contain cells not normally present in the tissue. The nuclear features of malignancy are anisokaryosis; pleomorphism; a high or variable N/C ratio; increased mitotic activity; nucleoli that vary in size, shape, and number; coarse chromatin; nuclear molding; and multinucleation (Fig. 10). Anisokaryosis is defined as a variation in nuclear size of 1.5 times or greater among cells in a population. Pleomorphism is variability in nuclear or cell shape. A pleomorphic population of neoplastic cells often has nuclei that are angular, elongated, or lobular. A high or variable N/C ratio is considered abnormal in most cell populations. The exception would be lymphocytes, which normally have a high N/C ratio. In most non-malignant cell populations, the N/C ratio is consistent from cell to cell and the nucleus rarely occupies more than one half of the cytoplasm (N/C ratio of 1:2). The identification of mitotic figures would indicate a rapidly dividing cell population and is used as a criterion for malignancy. On occasion, however, mitotic figures can be seen in normal or reactive tissues in which cell production is normally high. Such tissues include lymphoid tissue and bone marrow or other hematopoietic tissues. Many malignant cell populations contain nuclei with prominent and variable nucleoli. Because nucleoli can be seen in certain normal populations of cells (eg, hepatocytes), this criterion for malignancy is used when there is variability in the size, shape, or number of nucleoli in the cell population. A coarse chromatin or clumped chromatin pattern in the nucleus usually indicates an immature or extremely active cell population and is also used as a criterion for malignancy. Here, the nucleus contains dark heterochromatin (inactive DNA) and lighter and paler staining euchromatin (active DNA). Nuclear molding is when the nucleus of a cell compresses and deforms the shape of another nucleus in the same cell or an adjacent cell. This is a nuclear feature found in malignant cell populations that have lost contact inhibition for cell division and are growing in an uncontrolled fashion. Finally, the identification of multinucleated cells is sometimes used as a criterion for malignancy, particularly if the nuclei within the cell have malignant features, such as anisokaryosis, clumped chromatin, and prominent variable nucleoli. Some nonneoplastic cells may be multinucleated, particularly macrophages and osteoclasts (Fig. 11). Therefore, this criterion has to be interpreted with caution in the presence of pyogranulomatous inflammation or lytic bone lesions.

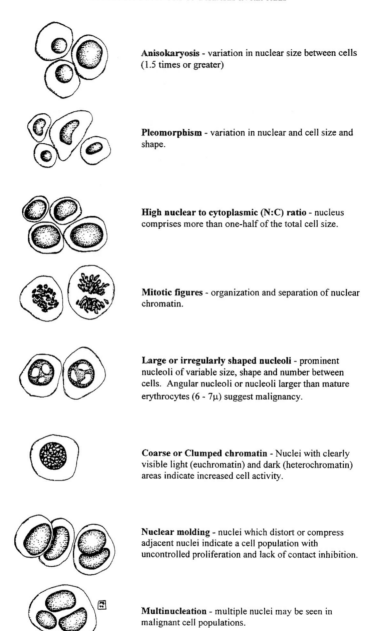

Fig. 10. Sketch of the various nuclear features that may be seen in a malignant neoplasm. (Courtesy of P.J. Bain, DVM, PhD, DACVP, North Grafton, MA).

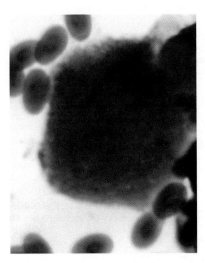

Fig. 11. Tissue aspirate from a lytic lesion in a tegu shows a large multinucleated osteoclast (Wright-Giemsa, original magnification ×50).

Typically, if three or more of these described malignant criteria are identified in a cell population, the tumor is malignant [1]. If only one or two malignant features are recognized in a cell population, malignancy may be suspected, but the diagnosis cannot be confirmed by cytology. In addition, it is important to recognize that the criteria for malignancy are not reliable if inflammation is present. Inflammation can cause reactive changes in epithelial or mesenchymal cells that mimic malignancy [1]. Therefore, if only a few cells display malignant characteristics or if inflammation is present, histopathologic examination is needed to confirm the diagnosis.

In contrast to nuclear features of malignancy, the cytoplasmic features and orientation of cells in a cell population are evaluated to determine the tissue of origin of the neoplastic cells. The four broad categories of neoplasia, with regard to tissue of origin, are epithelial, mesenchymal, round cell, and neuroendocrine. Fig. 12 illustrates the cytoplasmic features and orientation of the cells in the four tissue types.

Epithelial neoplasms are typically composed of large cells exfoliating in cohesive clumps or sheets. The cells have distinct cell borders and, typically, round or polygonal nuclei. The cytoplasmic membranes of adjacent cells form tight cytoplasmic junctions, which might appear as clear lines between cell cytoplasms. Aspirates taken from mesenchymal neoplasms are usually poorly cellular, with small to medium-sized spindle cells arranged individually or in loosely cohesive clusters (Figs. 13 and 14). The cytoplasm is often wispy, with trails of cytoplasm narrowing off in one or more directions out from the nucleus. Unlike epithelial cells, mesenchymal cells often have indistinct cell borders. The nuclei of mesenchymal cells are typically oval to polygonal.

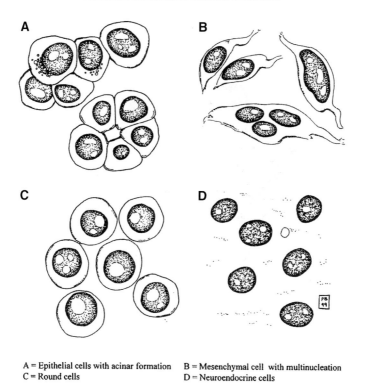

A = Epithelial cells with acinar formation   B = Mesenchymal cell with multinucleation
C = Round cells                              D = Neuroendocrine cells

Fig. 12. Sketch of the features of various tissue types that may be seen in a neoplastic process. (Courtesy of P.J. Bain, DVM, PhD, DACVP, North Grafton, MA.)

The cytologic features of round cell tumors include round to oval and small to medium individualized cells that exfoliate in moderate numbers. There are no cell-to-cell junctions, and the cells have discrete cell borders. These tumors are also described as discrete cell tumors [1]. Of the round cell tumors, histiocytic tumors, lymphomas, plasma cell tumors, and melanomas have been described in reptiles [6]. Lymphoma is probably the most frequently observed round cell tumor of reptiles. These tumors are composed of individually arranged round cells with deeply basophilic cytoplasm (Fig. 15). The nuclei are typically round, with a diffusely clumped chromatin pattern and visible nucleoli. Mitotic figures are occasionally seen. Because neoplastic lymphocytes are often large and fragile, many tissue aspirate preparations contain lysed cells in which only the pale-staining nucleus is visible. In addition, cytoplasmic fragments called lymphoglandular bodies are often observed in the background of the preparations [7].

Neuroendocrine neoplasms are tumors of chemoreceptors or endocrine glands. As a group, aspirates taken from these lesions share a common characteristic cytologic feature. The slide preparations appear as free or "naked" nuclei embedded in a background of cytoplasm, with few, if any, distinct

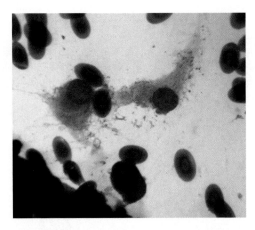

Fig. 13. Tissue aspirate taken from an osteosarcoma in a tegu. Two large spindle-shaped cells are seen, indicating a neoplasm of mesenchymal origin (sarcoma). The granules in the cytoplasm are intracytoplasmic osteoid (Wright-Giemsa, original magnification ×50).

cytoplasmic borders visualized [8]. Examples of commonly aspirated neuroendocrine tumors would be tumors of the thyroid gland and tumors of the adrenal gland.

## Mixed cell population

The mixed cell population is a category of lesions in which the preparation contains an inflammatory cell population (heterophils or macrophages) and a noninflammatory cell population (epithelial cells or mesenchymal cells). If this type of lesion is observed, a definitive cytologic diagnosis is

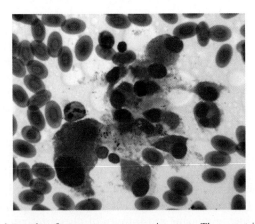

Fig. 14. Tissue aspirate taken from an osteosarcoma in a tegu. The eccentrically located nuclei and pink extracellular matrix (osteoid) are characteristic of osteosarcomas. Note the marked anisokaryosis, variable N/C ratio and multiple nucleoli (large cell) indicating a malignant process (Wright-Giemsa, original magnification ×50).

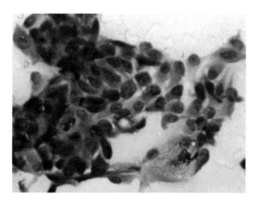

Fig. 15. Body cavity effusion from a leopard gecko contains a sheet of normal-appearing mesothelial cells. Note the pale basophilic cytoplasm, cytoplasmic vacuoles, round nuclei, and small nucleoli (Wright-Giemsa, original magnification ×20).

difficult to make. Inflammation can cause reactive changes in mesenchymal or epithelial cells that closely mimic malignancy. Therefore, it is nearly impossible to determine cytologically if the lesion is primarily an inflammatory response with reactive epithelium or mesenchymal cells or a neoplastic process that has become inflamed. In these circumstances, two approaches can be considered: treating the accompanying inflammation and reaspirating the lesion at a later date or performing an excisional or incisional biopsy for histologic evaluation [1].

## Fluid cytology

### Body cavity effusions

In most reptiles, body cavity or coelomic effusions may be identified and collected by paralumbar or paramidline percutaneous centesis [3]. In chelonians, the abdominal cavity is typically entered cranial to the hind limb. The coelomic cavity of normal reptiles contains scant fluid, and any fluid obtained should be examined. The position of the needle should be maintained parallel to the body wall to avoid enterocentesis, and the bevel should face toward the body wall to avoid aspiration of abdominal fat.

The complete laboratory evaluation of fluid includes total protein measurement, cell counts, and microscopic examination to perform a differential leukocyte count and to evaluate morphologic features of the cells. A refractometer is usually adequate for protein measurement in most fluids. The leukocyte count may be obtained by two methods: estimation from the smear or a manual cell count with a hemocytometer. Estimation from a smear is a crude method but allows approximations of total leukocyte counts until a more accurate determination can be made. Estimates are made using the following formula: (average number of leukocytes per microscopic field) × (objective power)² [9]. For example, if while using an original magnification ×20 objective, an average of 10 leukocytes per field is observed, the leukocyte count would be 10 × (20)², or 4000 cells per microliter. Microscopic examination of the fluid is done by smearing a single drop of collected fluid on a microscope slide and staining it as described in the sample preparation section. If the fluid is of low cellularity (<500 cells per microliter), it should by concentrated by centrifugation; the supernatant is then removed, and a drop of sediment can be used to make a smear for staining and evaluation. If the volume of fluid is insufficient to perform a complete analysis, a drop of fluid should be smeared on a microscope slide for staining, microscopic evaluation of cell types present, and estimation of the leukocyte count.

Coelomic fluids are classified as transudate, modified transudate, exudate, or neoplastic effusion (Table 2). Further subclassification within these major categories is sometimes necessary; for example, the identification of microorganisms or the absence of them would allow further classification of an exudate as a septic or nonseptic exudate, respectively.

Certain cell types are commonly encountered in effusions in reptiles, regardless of the type of effusion. Mesothelial cells that line the body cavity are often present in effusions. They may appear as individual round cells or in cohesive sheets of epithelium (see Fig. 15). These cells have a pale basophilic cytoplasm that may contain vacuoles. If they become reactive, the

Table 2
Cytologic parameters used in the classification of effusions [data from 3]

| Fluid type | Specific gravity | Protein (g/dL) | Cells/μL | Cell types present |
|---|---|---|---|---|
| Transudate | <1.020 | <3.0 | <1000 | Macrophages, small numbers of mesothelium, lymphocytes, and nondegenerate heterophils |
| Modified transudate | >1.020 | 3.0–3.5 | 1000–5000 | Macrophages, small numbers of mesothelium, lymphocytes, and heterophils |
| Nonseptic exudate | >1.020 | >3.0 | >5000 | Mixed population of heterophils (mostly nondegenerate), macrophages, and some lymphocytes |
| Septic exudate | >1.020 | >3.0 | >5000 | Mixed population of heterophils (mostly nondegenerate), macrophages, and some lymphocytes; microorganisms identified, preferably within phagocytes |
| Hemorrhagic | >1.020 | >3.0 | Similar to blood | Erythrocytes and peripheral blood leukocytes and macrophages, erythrophagia or hemosiderin in macrophages |
| Neoplastic | | Variable | Variable | Tumor cells, neoplastic cell population identified |

*Data from* Campbell TW. Clinical pathology. In: Mader DR, editor. Reptile medicine and surgery. Philadelphia: WB Saunders; 1996. p. 256.

cytoplasm can become more basophilic and a hair-like corona may be seen surrounding the cell. The nuclei are round to oval and usually contain a small nucleolus. Reactive mesothelium may have pleomorphic nuclei and other cytologic features that mimic malignancy [9]. A diagnosis of mesothelioma should not be made based on cytologic features of these cells in an effusion. Peculiar to reptiles, melanocytes are commonly observed in effusions. These cells can be quite large and characteristically contain dark, rod-shaped, green-black cytoplasmic granules (Fig. 16). This pigment is often refractile when brought in and out of focus during microscopic examination. Care should be taken so as not to confuse melanin pigment with bacterial agents. Bacteria are not refractile and stain dark blue in preparations stained with Romanowsky-type stains.

As in mammalian species, pure transudates are most frequently a result of hypoproteinemia and associated loss of normal plasma oncotic pressure within the vascular system [9]. Decreased production of albumin, increased loss, or a relative decrease in plasma albumin may potentially lead to a pure transudate effusion. Transudates are typically transparent straw-colored fluids. They are, by definition, fluids of low protein ($<3.0$ g/dL) and leukocyte counts ($<1000$ cells per microliter) [3]. A direct preparation of a transudate is a poorly cellular sample, with a colorless or lightly eosinophilic background containing few contaminant erythrocytes and, typically, small numbers of macrophages or heterophils. Variable numbers of mesothelial cells may be seen along with melanocytes or macrophages containing melanin pigment.

A modified transudate is an effusion that occurs by transudative mechanisms in which vascular fluids leak out of "normal" or "noninflamed" vessels (eg, by increased capillary hydrostatic pressure or lymphatic obstruction) [9]. Leakage of fluid from these vessels carries high protein content and is thus

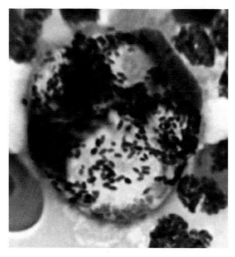

Fig. 16. Body cavity effusion from a python contains a melanocyte with dark, rod-shaped, intracytoplasmic pigment granules (Wright-Giemsa, original magnification ×100).

"modified" by the addition of protein or cells. Grossly, modified transudates may vary from straw colored, to slightly turbid, to pink, to white depending on the cause. The protein content ranges from 3.0 to 3.5 g/dL, and the cell count is between 1000 and 5000 cells per microliter [3]. Microscopic evaluation of a direct preparation typically consists of a lightly basophilic background, with variable numbers of moderately to markedly vacuolated reactive macrophages, mesothelial cells, and a few small lymphocytes or heterophils (Fig. 17). As with transudates, melanocytes or melanophages may be seen.

In contrast to transudates and modified transudates, exudates result from leakage of fluid from abnormal or altered vasculature [9]. This generally occurs because of an inflammatory process or chemotactic stimuli within the coelomic cavity. The inflammatory process increases serosal and vascular permeability, resulting in a protein-rich fluid ($\geq 3.0$ g/dL) and increased leukocyte count ($\geq 5000$ cells per microliter) [3]. Exudates are further classified as septic or nonseptic depending on whether or not infectious agents are identified in the fluid.

Information regarding the cause and chronicity of exudates can be ascertained by evaluating the types of leukocytes present and any changes in normal cell morphology. Acute inflammatory processes typically consist predominantly of heterophils (Fig. 18). If bacterial or fungal agents are identified, particularly in phagocytic vacuoles within leukocytes, the exudate is classified as septic (Fig. 19). Other cytologic features associated with sepsis are degenerative changes of the heterophils, specifically karyorrhexis, karyolysis, and degranulation of heterophils.

Nonseptic exudates typically have mononuclear phagocytes as the predominant leukocyte, with a lesser population of heterophils. In nonseptic effusions, the leukocytes would have minimal, if any, degenerative changes, and microorganisms would not be identified. A common cause of septic or nonseptic

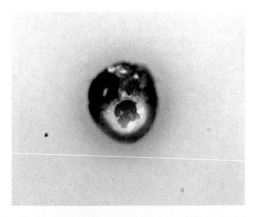

Fig. 17. Body cavity effusion from an iguana. This is a modified transudate with a basophilic background and a reactive macrophage containing a phagocytized leukocyte (leukopohagia) (Wright-Giemsa, original magnification ×100).

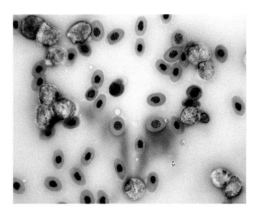

Fig. 18. Body cavity effusion from an iguana. An acute inflammatory exudate is seen with a predominant population of heterophils. Two macrophages are also noted in the upper left (Wright-Giemsa, original magnification ×50).

exudates is egg yolk peritonitis. Egg yolk peritonitis results from the yolk being released into the coelomic cavity. This causes a severe inflammatory response and is typically associated with a poor prognosis. Grossly, a variably thick opaque fibrinous fluid is present. The characteristic cytologic features are a highly proteinaceous basophilic background, with colorless variably sized lipid vacuoles, basophilic globules of proteinaceous material, and inflammatory cells (Fig. 20). Unlike many nonseptic exudates, egg yolk peritonitis typically involves an acute presentation and a predominantly heterophilic inflammatory response. Lesser numbers of moderately to markedly vacuolated reactive macrophages may be seen. Circular to polygonal basophilic structures of proteinaceous material are frequently identified in the background of the smear and within the cytoplasm of phagocytes (Fig. 21).

A hemorrhagic effusion is suspected when the fluid contains numerous erythrocytes and iatrogenic contamination of the sample has been ruled

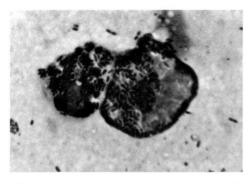

Fig. 19. Body cavity effusion from an iguana. A septic exudate is seen with two degenerate heterophils containing phagocytized intracellular bacteria (Wright-Giemsa, original magnification ×100).

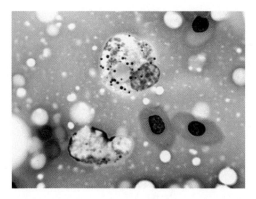

Fig. 20. Body cavity effusion from an iguana with egg yolk peritonitis. There is a densely basophilic background that contains clear lipid vacuoles and three globules of basophilic proteinaceous material (*bottom left*). One immature heterophil (*top*), one mature heterophil (*bottom*), and three erythrocytes (*right*) are seen (Wright-Giemsa, original magnification ×100).

out. Hemorrhagic effusions may result from ruptured vessels or alterations in vascular endothelial integrity. Hemorrhagic effusions grossly and microscopically contain a certain amount of blood, and the packed cell volume should be at least 10% to 25% of the peripheral blood for a fluid to be classified as hemorrhagic [9]. There are no specific numeric values that define a hemorrhagic effusion; however, hemorrhagic fluid with leukocyte counts higher than that seen in the peripheral blood should be considered inflammatory as well.

Several factors help in distinguishing between hemorrhagic effusion and iatrogenic blood contamination that might occur during sampling. One distinguishing factor is that thrombocytes are usually not seen in hemorrhagic effusions present for more than 1 hour before sample collection. Hemorrhagic

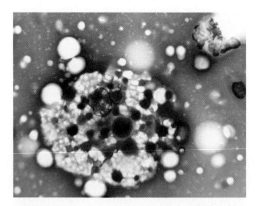

Fig. 21. Body cavity effusion from an iguana with egg yolk peritonitis. A densely basophilic background that contains clear lipid vacuoles and a single large macrophage with phagocytized basophilic globules is present. One heterophil is also seen in the upper right (Wright-Giemsa, original magnification ×100).

effusions also contain reactive macrophages with phagocytized erythrocytes or intracytoplasmic hemosiderin or hematoidin (iron pigments) (see Fig. 9). Additionally, blood that is the result of hemorrhage into a body cavity does not clot even in a clot tube. Conversely, iatrogenic contamination with peripheral blood during sampling contains thrombocytes, does not have erythrophagocytosis, and usually clots after collection. The packed cell volume of this fluid is equal to that of the peripheral blood if a vessel is punctured. In peracute hemorrhage occurring less than 45 minutes before sampling, it may be difficult to distinguish iatrogenic contamination from a hemorrhagic process [9].

Common causes of hemorrhagic effusions are trauma and injury, particularly to vascular organs. Neoplastic processes must also be considered as potential causes for hemorrhage into the coelomic cavity, however [3]. Rupture or leakage of vessels within a tumor may result in the accumulation of a hemorrhagic effusion. Coagulopathies may also result in the accumulation of blood in the coelomic cavity and may be considered if no evidence of trauma or neoplasia is present and the clinical history or physical examination suggests coagulopathy.

The term *neoplastic effusion* should only be used to describe effusions in which a neoplastic cell population has been identified in the fluid. Making this determination may be difficult, because neoplastic cells are often not present in effusions caused by tumors and because reactive mesothelial cells can have cytologic criteria that mimic malignancy [9]. There are no particular numeric parameters for neoplastic effusion; however, the total protein content is typically high ($>3.0$ g/dL). These fluids are often inflammatory and may also have evidence of hemorrhage. As seen in mammalian species, carcinomas, mesotheliomas, and round cell tumors are the neoplastic processes most likely to result in an effusion that contains a neoplastic cell population [9]. In carcinomas that result in a neoplastic effusion, it may be difficult to distinguish the malignant cell population from a population of reactive macrophages or mesothelial cells. This distinction is frequently challenging because of the bizarre features frequently seen in reactive cell populations. Cytologic features that may help to distinguish reactive mesothelium and macrophages from carcinoma cells include the following malignant criteria: (1) marked pleomorphism; (2) macrokaryosis (giant nuclei); (3) large, angular, or multiple nucleoli; (4) multinucleation; and (5) increased mitotic activity [9]. Lymphoma may result in a neoplastic effusion in which the fluid contains a homogeneous population of large lymphoblasts with moderate amounts of basophilic cytoplasm and large nuclei, often two to four times the size of a small well-differentiated lymphocyte.

*Synovial fluid*

Synovial fluid may be obtained and evaluated in any patient suspected of having an arthropathy. Sterile preparation of skin over the joint of interest followed by penetration of the joint capsule with a fine needle and gentle aspiration of synovial fluid may yield a diagnostic sample. Evaluation of synovial fluid includes the assessment of color, volume, viscosity, protein content, total cell count, and differential cell count of the fluid. As with fluids collected from animals with effusions, if only a small amount of fluid is collected, the drop should be smeared on a slide, stained, and examined microscopically. Based on this microscopic evaluation, estimates of leukocyte numbers and quality of mucin as well as cytologic evaluation can be performed.

Classification of synovial fluid as inflammatory or noninflammatory is made based on the mucin quality, leukocyte count, differential, and cytologic appearance of the leukocytes. Viscosity (mucin quality) may be assessed on gross examination by placing a drop of synovial fluid onto a clean glass slide [10]. Normal synovial fluid is extremely viscous because of the high concentration of hyaluronic acid. When the needle or syringe is separated from a droplet of fluid that is placed on a glass slide, the strand formed should reach a length of 2 cm or greater before separating from the needle. If the strand breaks at a length of less than 2 cm, the viscosity of the joint is considered to be reduced.

Normal synovial fluid is viscous and poorly cellular, containing a densely eosinophilic background (mucin), few mononuclear leukocytes, and virtually no erythrocytes unless associated with capillary bleeding from sample collection (Fig. 22). Occasionally, cells within synovial fluid are seen in a linear arrangement known as "windrowing" (Fig. 23). This may suggest that the fluid has good viscosity, and it can be seen in any fluid that contains large numbers of leukocytes or erythrocytes [10].

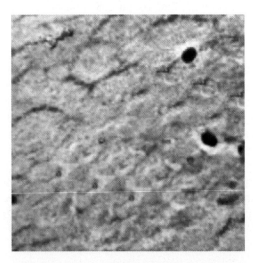

Fig. 22. Normal synovial fluid from a gopher tortoise with a densely eosinophilic granular background indicates normal viscosity and mucin content. Two small mononuclear cells are seen on the right (Wright-Giemsa, original magnification ×50).

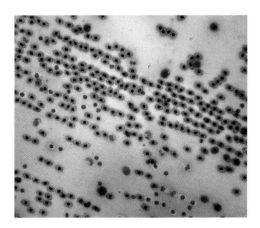

Fig. 23. Blood-contaminated synovial fluid from a green iguana. Notice the linear arrangements of the numerous erythrocytes (windrowing). A small number of heterophils and few large mononuclear cells are also present (Wright-Giemsa, original magnification ×20).

Inflammatory joint fluid contains an increased number of heterophils beyond what might be expected from any blood contamination, along with an increased number of large mononuclear cells (Fig. 24). Inflammation within the synovial fluid warrants culture of the fluid and close microscopic examination for microorganisms. If microorganisms are identified microscopically, particularly within the cytoplasm of phagocytes, a diagnosis of septic arthritis is made.

In reptiles, noninflammatory joint fluid is often the result of mineral deposits within the joint, which can occur as a result of aging, dietary imbalances, chronic drug administration, or chronic renal disease [5,11]. Mineral

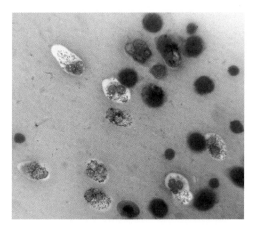

Fig. 24. Synovial fluid from a lizard with inflammatory joint disease. The fluid contains several heterophils (*orange granules*) and basophilic large mononuclear cells. A single erythrocyte is seen at bottom center (Wright-Giemsa, original magnification ×100).

deposits may consist of uric acid crystals, as seen in gout, or calcium deposits, as seen in pseudogout [5,11]. Gout may affect visceral organs and joints [11]. Synovial fluid from animals with articular gout is thick, opaque, and chalky white [12]. Microscopically, the fluid contains numerous brown needle-like crystals arranged individually and in small starburst-like aggregates (Fig. 25). Synovial fluid from reptiles (typically lizards) with pseudogout contains small to moderate numbers of reactive macrophages and round to irregularly shaped refractile crystalline material (Fig. 26). These calcium-containing crystals stain poorly with Wright-Giemsa but stain dark brown to black using a calcium-identifying stain, such as Von Kossa's stain.

**Respiratory cytology**

Cytologic specimens used to evaluate the respiratory tract include nasal, oral, or tracheal swabs; transtracheal washes; or percutaneous lung washes. Proper procedures for collection of these samples are covered elsewhere [13]. Samples collected from tracheal or percutaneous lung washes may be used for cytologic evaluation as well as microbial culture for the identification of microorganisms. Direct smears or sedimented samples may be prepared and stained for examination of cells or detection of parasites, bacterial elements, or fungal elements. Slides should be prepared quickly after collection to avoid cellular swelling and lysis in the saline. Any flocculent or solid material identified in the sample should be selectively secured with a capillary tube, pipette, or sterile applicator and placed on a clean glass slide. The material should be spread gently between two slides as described in the section on sample collection and slide preparation. Mucus clusters often contain important diagnostic material, such as leukocytes and microorganisms. Direct smears of fluid sediment are similarly prepared.

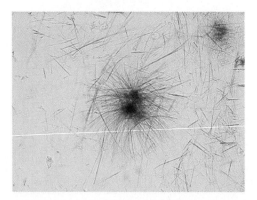

Fig. 25. Synovial fluid from an African spur-thighed tortoise with gout. Tan needle-like uric acid crystals are seen individually and in starburst-like configurations (unstained, original magnification ×50). (Courtesy of R.E. Raskin, DVM, PhD, DACVP and A. Casimire-Etzioni).

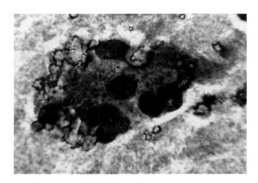

Fig. 26. Synovial fluid from a veiled chameleon with pseudogout. A large multinucleated macrophage is seen surrounded by poorly staining refractile calcium deposits (Wright-Giemsa, original magnification ×100). (Courtesy of H.L.Wamsley, DVM, DACVP, Gainesville, FL.)

Oropharyngeal or nasal swabs normally contain superficial squamous epithelial cells and, in the case of oral swabs, surface bacteria. Squamous epithelium cells are relatively large, with abundant angular cytoplasm and a small round nucleus. Bacteria are often seen adhered to the surface of squamous epithelial cells or present in the background of the smear. This does not indicate a septic process unless heterophils containing phagocytized microorganisms are present.

The normal components of the respiratory tract include a small amount of mucus, ciliated columnar epithelial cells, goblet cells, and alveolar macrophages. Leukocytes are rarely present in normal tracheal or lung washes. Mucus in small amounts is normal and may appear as lightly basophilic or eosinophilic amorphous material, frequently in swirls (Fig. 27). Ciliated columnar epithelial cells are elongated or cone-shaped cells with pink cilia on one end and a round or oval nucleus (Fig. 28). Goblet cells are mucus-producing bronchial cells seen in small numbers in normal tracheal washes.

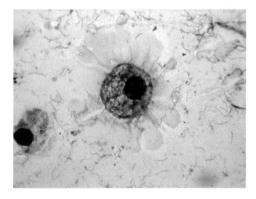

Fig. 27. Transtracheal wash from a normal alligator. An intact large alveolar macrophage (*center*) and a smaller lysed macrophage are seen. The background is stippled with mucous strands (Wright-Giemsa, original magnification ×100).

Fig. 28. Transtracheal wash from a rat snake contains two ciliated columnar epithelia. These are a normal cellular component of the respiratory tract (Wright-Giemsa, original magnification ×100).

They are nonciliated columnar cells with a basally placed nucleus and cytoplasm that is distended with large round red or blue mucin granules. Occasionally, mucin granules may be noted free in the background of some smears. Alveolar macrophages are the predominant cell type seen in normal tracheal washes. These cells are large and round, with a pale blue and sometimes vacuolated cytoplasm (see Fig. 27). The nucleus is round to reniform in shape and eccentrically located in the cytoplasm. When these macrophages become activated, the cytoplasm becomes more vacuolated and may contain phagocytized material.

Microorganisms are not a normal component of the respiratory tract and are only visualized if there is a septic process or if oral contamination occurred during the sampling procedure. Oral contamination is more likely when collection is by the tracheal wash technique than when collection is by way of the percutaneous lung. It is important to be able to recognize oral contamination so that it is not mistaken for a septic process in the lungs. The hallmark of oral contamination is the presence of superficial squamous epithelium, some of which contains a heterogeneous surface bacterial population. The presence of squamous epithelial cells and bacteria in the absence of an inflammatory response is consistent with oral contamination.

Cytologic evaluation of the respiratory tract is primarily used to identify an inflammatory or infectious process. The presence of increased amounts of mucus or heterophils and increased numbers or reactive macrophages would indicate an inflammatory process. The identification of microorganisms accompanying an inflammatory process would indicate sepsis (Figs. 29 and 30).

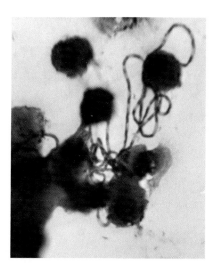

Fig. 29. Transtracheal wash from a corn snake with bacterial pneumonia. Numerous heterophils are present along with filamentous bacterial organisms (*center*) (Wright-Giemsa, original magnification ×50).

## Gastrointestinal cytology

Cytologic specimens used to evaluate the gastrointestinal tract include oral swabs, gastric lavage, cloacal swabs, and cloacal washes. These types of samples are typically evaluated to obtain information about the bacterial flora as well as to identify any fungal, protozoal, or parasitic pathogens. The

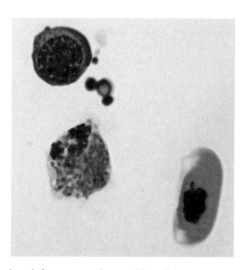

Fig. 30. Transtracheal wash from a green iguana. The cell types present are a reactive lymphocyte (*top left*), a heterophil (*center*), and an erythrocyte (*bottom right*). A basophilic budding yeast organism (between lymphocyte and heterophil) consistent with *Candida* spp was identified throughout the preparation (Wright-Giemsa, original magnification ×100).

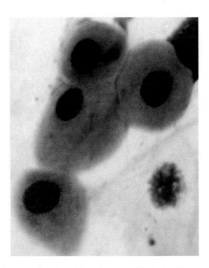

Fig. 31. Gastric lavage from a tortoise. Four large epithelial cells are noted. These are part of the normal cell population (Wright-Giemsa, original magnification ×100).

resident epithelial cell population depends on the specific area being sampled. The oral cavity is lined by superficial squamous epithelium, and the gastrointestinal tract contains cuboidal to columnar secretory epithelial cells. Tissue aspiration cytology can be performed on neoplastic cells that may arise from these epithelial cell populations and have the same features of neoplasia or malignancy that were described previously in this article. Common epithelial tumors of the oral cavity and gastrointestinal tract include papillomas, squamous cell carcinomas, adenomas, and adenocarcinomas [6]. Sarcomas have also been reported in the oral cavity (fibrosarcoma and rhabdomyosarcoma) and gastrointestinal tract (leiomyosarcomas) of reptiles [6]. The most common round cell tumor observed in

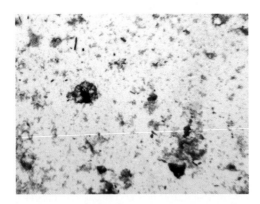

Fig. 32. Gastric lavage from an emerald tree boa. There is basophilic staining mucus in the background and a heterogeneous population of normal bacterial flora (rod-shaped organisms of various sizes) (Wright-Giemsa, original magnification ×50).

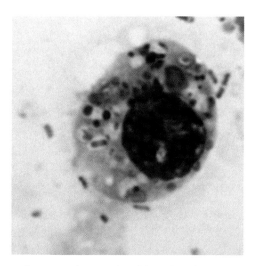

Fig. 33. Gastric lavage from a gopher tortoise with septic inflammation. The heterophil contains phagocytized intracytoplasmic bacterial rods (Wright-Giemsa, original magnification ×100).

the gastrointestinal tract of reptiles is lymphoma [6]. The cytologic features of these lesions would be the same as described previously in this article.

Oral swabs typically contain epithelial cells, a heterogeneous bacterial population, and few, if any, leukocytes. Inflammatory lesions are characterized by the presence of leukocytes, typically heterophils. Plant material may also be present as a contaminant in oral swabs.

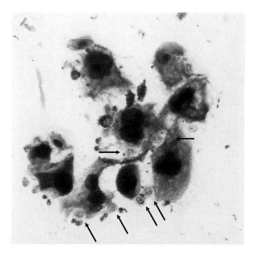

Fig. 34. Gastric lavage from a rat snake with cryptosporidiosis. Numerous *Cryptosporidium* oocysts are seen (*arrows*) among a group of variably sized reactive gastric epithelium with basophilic cytoplasm and binucleation (Wright-Giemsa, original magnification ×50).

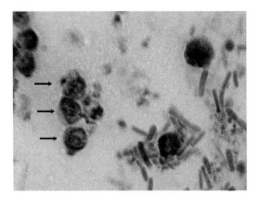

Fig. 35. Gastric lavage from a rat snake with cryptosporidiosis. Numerous *Cryptosporidium* oocysts (*arrows*) are identified as positive (acid-fast stain, original magnification ×100).

Gastric lavage samples normally contain a heterogeneous bacterial population, epithelial cells, a few leukocytes, and a small amount of mucus (Figs. 31 and 32). The presence of a monomorphic bacterial population may indicate overgrowth of an abnormal bacterial population. Increased numbers of inflammatory leukocytes (heterophils, lymphocytes, and macrophages) would suggest inflammatory bowel disease and warrant further investigation for the presence of parasites, protozoa, and fungal or bacterial pathogens (Fig. 33). *Cryptosporidium* oocysts may be identified in gastric lavage fluid (or fecal analysis) from affected reptiles, especially snakes. Some snakes can be subclinical carriers and may intermittently shed organisms [14]. In animals with clinical disease, the organism causes gastric epithelial

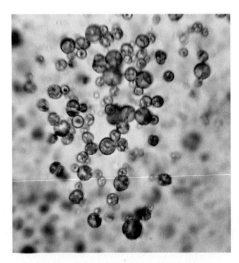

Fig. 36. Cloacal swab from a boa constrictor. The numerous radiant spheres of various sizes are urate crystals (unstained, original magnification ×50).

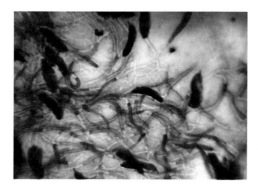

Fig. 37. Cloacal swab from a ball python. Numerous spear cells are identified (Wright-Giemsa, original magnification ×100).

hyperplasia and associated gastrointestinal signs, primarily regurgitation [14]. Diagnosis can be made by identification of small (6 μm) round to oval oocysts in stained preparations of gastric lavage fluid or fecal swabs. The organisms stain lightly basophilic, with a stippled cytoplasm, and contain a clear thin capsule surrounding the oocyst (Fig. 34). Acid-fast stain has been used to accentuate the organism and aid in identification. The organisms stain an orange red with acid-fast stain (Fig. 35).

Similar to gastric lavage, cloacal swabs or washes can also be used to identify inflammatory disease or infectious or parasitic organisms. Cytologic preparations of swabs taken from normal reptiles typically contain epithelial cells, a small amount of mucus, urate crystals, and, sometimes, a heterogeneous bacterial population (Fig. 36). A large number of sperm cells can be seen in cloacal swabs from male reptiles (Fig. 37). Leukocytes should be few in number or absent in cloacal swabs taken from normal animals but may be seen in increased numbers in animals with inflammatory or infectious diseases.

**Summary**

The evaluation of cytologic preparations can be a useful tool in the identification of various disease processes in many different reptile species. Although the class Reptilia contains numerous species of animals, the basic cytologic features of the various disease processes apply to all. In addition, the basic categories of pathologic processes (eg, inflammation, neoplasia) can be applied to samples taken from many tissue masses and internal organs. Sometimes, a specific diagnosis can be made from a cytologic evaluation. In situations in which a specific diagnosis cannot be made, however, information obtained from a thorough cytologic evaluation may be used to help determine the correct course of action.

## References

[1] Tyler RD, Cowell RL, Baldwin CJ, et al. Introduction. In: Cowell RL, Tyler RD, editors. Diagnostic cytology of the dog and cat. Goleta (CA): American Veterinary Publications; 1993. p. 1–19.
[2] Meyer DJ. The acquisition and management of cytology specimens. In: Raskin RE, Meyer DJ, editors. Atlas of canine and feline cytology. Philadelphia: WB Saunders; 2001. p. 1–18.
[3] Campbell TW. Clinical pathology. In: Mader DR, editor. Reptile medicine and surgery. Philadelphia: WB Saunders; 1996. p. 248–57.
[4] Rossi JV. Dermatology. In: Mader DR, editor. Reptile medicine and surgery. Philadelphia: WB Saunders; 1996. p. 104–17.
[5] Barten SL. Lizards. In: Mader DR, editor. Reptile medicine and surgery. Philadelphia: WB Saunders; 1996. p. 324–32.
[6] Done LB. Neoplasia. In: Mader DR, editor. Reptile medicine and surgery. Philadelphia: WB Saunders; 1996. p. 125–41.
[7] Raskin RE. Lymphoid system. In: Raskin RE, Meyer DJ, editors. Atlas of canine and feline cytology. Philadelphia: WB Saunders; 2001. p. 93–134.
[8] Alleman AR. Endocrine system. In: Raskin RE, Meyer DJ, editors. Atlas of canine and feline cytology. Philadelphia: WB Saunders; 2001. p. 385–400.
[9] Alleman AR. Abdominal, thoracic and pericardial effusions. Vet Clin North Am Small Anim Pract 2003;33:89–118.
[10] Parry BW. Synovial fluid. In: Cowell RL, Tyler RD, editors. Diagnostic cytology of the dog and cat. Goleta (CA): American Veterinary Publications; 1993. p. 121–36.
[11] Mader DR. Gout. In: Mader DR, editor. Reptile medicine and surgery. Philadelphia: WB Saunders; 1996. p. 374–9.
[12] Casimire-Etzioni AL, Wellehan JF, Embury JE, et al. Synovial fluid from an African spur-thighed tortoise (Geochelone sulcata). Vet Clin Pathol 2004;33(1):43–6.
[13] Jenkins JR. Diagnostic and clinical techniques. In: Mader DR, editor. Reptile medicine and surgery. Philadelphia: WB Saunders; 1996. p. 264–76.
[14] Cranfield MR, Graczyk TK. Cryptosporidiosis. In: Mader DR, editor. Reptile medicine and surgery. Philadelphia: WB Saunders; 1996. p. 359–63.

# Cytologic Diagnosis of Disease in Amphibians

Allan P. Pessier, DVM, DACVP[a]

[a]*Wildlife Disease Laboratories, Conservation and Research for Endangered Species, Zoological Society of San Diego, PO Box 120551, San Diego, CA 92112-0551, USA*

Cytology is an inexpensive yet powerful diagnostic tool that allows for rapid diagnosis of many common disease conditions in amphibian patients. Although the emphasis of this article is on infectious diseases, there is great potential for application of cytologic diagnosis to variety of medical conditions as the knowledge base in amphibian medicine and pathology continues to grow. Routine methods used that may fall under the umbrella of cytology range from wet mount examination of skin scrapings (or gill biopsies of larvae) to examination of stained impression smears. Routine Romanowsky's-type stains (eg, Wright-Giemsa) work well for amphibian samples. Preparation of multiple smears is always recommended to allow for use of special staining procedures, such as Gram stain and stains for acid-fast bacilli.

## Peripheral blood

A complete discussion of amphibian hematology is beyond the scope of this article and a clinically oriented review is available [1]. A few aspects of blood cell morphology that are important in the review, discussion, and interpretation of clinical cytology specimens are mentioned here.

### Erythrocytes

Most amphibian erythrocytes are elliptical and nucleated and are among the largest observed in the vertebrates (up to 10–70 µm in diameter). In some species there are varying numbers of anucleate erythrocytes and this feature is especially prominent in the lungless (plethodontid) salamanders, presumptively to increase surface area for oxygen transport. Immature erythrocytes

*E-mail address:* apessier@sandiegozoo.org

(erythroblasts) and erythrocyte mitoses may be observed in blood films and may be normal or of uncertain clinical significance [1–3].

*Leukocytes*

The terminology used to describe amphibian leukocytes can be confusing and inconsistent between different veterinarians, veterinary pathologists, and laboratory technicians. Although extrapolation of information from that described in other vertebrates, such as reptiles, may be useful, it is also important to remain cognizant that the Amphibia are a distinct and diverse class of animals for which such assumptions may be easily erroneous.

Three types of granulocyte have been described in amphibians [1–4]. The first is compared with the mammalian neutrophil and has variably segmented nuclei, indistinct cytoplasm, and usually lacks distinct cytoplasmic granules (Fig. 1). The second type of granulocyte (eosinophilic granulocyte) is characterized by abundant orange to red cytoplasmic granules. Cells termed "coarse" eosinophilic granulocytes have large round granules resembling avian or mammalian eosinophils, whereas "fine" eosinophilic granulocytes have small rod to brick-shaped granules that resemble avian heterophils. The third type of granulocyte is the basophil with abundant

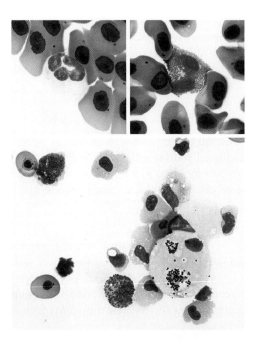

Fig. 1. A neutrophil (*top left*) and coarse eosinophilic granulocyte (*top right*) from a fire salamander (*Salamandra salamandra*). The bottom photomicrograph shows a basophil (*bottom middle*) and coarse eosinophilic granulocyte (*top left*) in coelomic fluid from a Surinam toad (*Pipa pipa*). There is also a large melanomacrophage (*bottom right*).

purple cytoplasmic granules that often obscure the nucleus. Differences between cells have been based predominantly on morphology with little in the way of functional or cytochemical study and caution should be used in equating morphologic appearance with homologous cellular function to similar cells in other vertebrates. For instance, the mammalian eosinophil is frequently observed with parasitism or type-I hypersensitivity reactions, whereas establishment of similar activity has been inconclusive for other nonmammalian vertebrates [5].

*Erythrocytic inclusions and hemoparasites*

Intracytoplasmic inclusions, vacuoles, or areas of cytoplasmic clearing are commonly observed in amphibian erythrocytes from both peripheral blood films and specimens for diagnostic cytology. Differentials include sample collection or slide preparation artifact, nonspecific cytoplasmic accumulation of degenerate organelles, bacterial or viral infection, and hemoparasites [6]. In most cases inclusions are incidental findings and definitive diagnosis is possible only by use of such techniques as transmission electron microscopy.

Erythrocytic cytoplasmic inclusions caused by bacterial organisms have been reported, but have no known clinical significance [7–9]. Inclusions associated with *Aegyptianella ranarum* or *Aegyptianella bacterifera* are described as translucent spherical to elliptical cytoplasmic vacuoles [7,8]. Although *A ranarum* was originally considered to be in the family Anaplasmataceae, order Rickettsiales, recent molecular evidence suggests it could be a member of the Flavobacteriaceae [10].

Icosahedral viral particles suggestive of iridoviruses are associated with erythrocytic cytoplasmic inclusions that are variably described as roughly spherical, trapezoidal, or quadrilateral with acidophilic to basophilic staining properties [11,12]. Other morphologic changes in erythrocytes that may be associated with the erythrocytic iridoviruses include spheroidal, rather than the normal elliptical erythrocyte shape; nuclear displacement; and cytoplasmic vacuoles or pale-staining areas (albuminoid vacuoles or bodies). Similar erythrocytic iridoviruses have been described in reptiles and the inclusions were once thought to represent protozoa in the genus *Pirhemocyton* [13]. In most cases erythrocytic iridovirus infections are incidental findings; however, anemia was noted in heavily infected bullfrogs (*Rana catesbeiana*) [12].

The erythrocytic iridoviruses are distinct from iridoviruses in the genus *Ranavirus* that are associated with mass mortality events in a variety of amphibians [14,15]. Ranavirus infections are multisystemic and histologically large basophilic intracytoplasmic inclusions may be observed in a variety of locations including hematopoietic tissue. Ranaviral inclusions have been observed in circulating leukocytes of an infected eastern box turtle and theoretically could also be seen in peripheral blood films or tissue impression smears of amphibians [16]. Icosahedral virions observed in the cytoplasm of leukocytes of ranid frogs (frog leukocytic virus) have not been fully characterized [17].

A wide variety of hemoparasites including nematodes (microfilaria), flagellate protozoa (trypanosomes and diplomonads), and apicomplexan protozoa are observed in amphibians and may be found incidentally in tissue impression smears [18,19]. Apicomplexan protozoa include the genera *Lankesterella*, *Schellackia*, *Babesiosoma*, *Dactylosoma*, and the hemogregarines, *Hemogregarina* and *Hepatozoon*. Most hemogregarines in amphibians have been placed into the genus *Hepatozoon* and organisms are usually observed in erythrocyte cytoplasm [20]. The elongate mature gamonts are observed parallel to the long axis of the erythrocyte and other morphologic changes, such as nuclear displacement, areas of cytoplasmic clearing, and megalocytosis of infected cells, may be observed (Fig. 2). Infections are usually incidental findings; however, very heavy infections may be clinically significant.

## Integument

Among the vertebrates the skin of amphibians has unique physiologic functions that can include water absorption, osmoregulation, and even respiration in some species. Reflecting these functions, the epidermis is thin and minimally keratinized without protective structures, such as hair or feathers. Other normal features that may be relevant to examination of cytologic preparations include a fine mineralized layer in the superficial dermis of some species (Eberth-Kastchenko layer) and the presence of chromatophores that contribute to cutaneous coloration (see section on chromatophore hyperplasia).

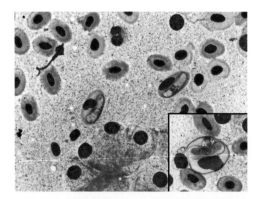

Fig. 2. A liver impression smear from a leopard frog (*Rana pipiens*). Two erythrocytes contain elongate mature gamonts of a *Hepatozoon* sp with displacement of their nuclei. The inset shows marked enlargement (megalocytosis) and cytoplasmic clearing in an infected erythrocyte. A cluster of hepatocytes (*bottom, middle*) have areas of cytoplasmic clearing suggestive of a glycogen-type vacuolar change.

The minimal protective features of amphibian skin results in increased susceptibility to cutaneous injury and disease, and skin lesions that are of minor importance to other vertebrates become significant because they may disrupt cutaneous physiologic function. As a result, evaluation of the integument is a critical component of the routine clinical and postmortem examination of amphibian patients. The spectrum of skin disease in amphibians has recently been reviewed [21,22]. Wet mount or cytologic examination of skin scrapings, impression smears (touch preparations) of cutaneous ulcers, fine-needle aspirates, and preparations of shed skin fragments can all provide important information to guide diagnosis and treatment.

Methods for obtaining clinical samples for wet mount or cytologic evaluation have been described in detail [23]. Wet mount examination of skin scrapings is often the best technique for detection of both protozoal and metazoan parasites and some fungal infections. Impression smears of cutaneous ulcers can be helpful in detecting infectious agents or characterizing bacterial populations in persistent lesions. Some caution in interpretation is necessary because the relatively superficial cellular sampling obtained in impression smears may not be diagnostic or provide misleading information, such as evidence of a bacterial infection in a more deeply situated primary fungal, mycobacterial, or neoplastic lesion. Fine-needle aspirates using the same techniques as described for domestic animals are used for discrete nodules and provide the most representative sampling of these lesions. Skin fragments shed into the environment and processed for cytologic examination can be a noninvasive diagnostic tool, especially for superficial fungal infections, such as chytridiomycosis. Recovered skin fragments are flattened onto glass slides, thoroughly air dried, and stained routinely for cytologic examination (Fig. 3). An important caveat to the use of this technique is

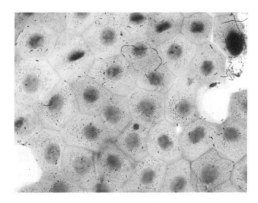

Fig. 3. A sheet of normal shed skin from a mountain yellow-legged frog (*Rana muscosa*) that has been dried flat on a glass slide and processed for cytologic examination. Individual epithelial cells are polygonal and many have central pink-red nuclei. There are small numbers of mixed bacteria.

that environmental bacteria and water molds can rapidly colonize shed skin and care should be taken not to overinterpret their presence.

*Bacterial skin disease*

Small to moderate numbers of predominantly gram-negative bacteria are frequently observed in cytologic preparations of normal amphibian skin and reflect common aquatic environmental flora including such organisms as *Aeromonas* and *Pseudomonas* sp, among many others. Bacterial skin diseases are common in captive animals and frequently occur secondary to trauma; exposure to environmental irritants; or as complications of other infectious diseases, such as cutaneous mycoses or parasitism. Furthermore, bacterial septicemia can have cutaneous manifestations, such as hyperemia, ulceration, subcutaneous edema, and hemorrhage, and these have been traditionally termed "red leg syndrome." In view of the normal bacterial skin flora, the observation of phagocytosed bacteria within granulocytes or macrophages is an important criterion for determining the presence of a true primary or secondary bacterial infection (Fig. 4). Serial cytologic examinations of cutaneous ulcers over time may provide helpful information on clinical response to therapy.

*Mycobacteriosis*

Infections by a variety of mycobacteria including *Mycobacterium marinum*, *M fortuitum*, *M xenopi*, and *M chelonei* are frequently observed in captive amphibians and may have both cutaneous and systemic manifestations [21,22,24]. Most are opportunistic pathogens and are frequently encountered in aquatic environments. The clinical presentation is variable and the spectrum of lesions includes discrete cutaneous nodules; ulcers; regional swelling (cellulitis); and abscesses. The cytologic appearance of mycobacterial

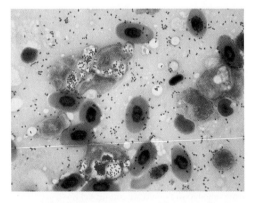

Fig. 4. Impression smear of a cutaneous ulcer in a golden mantella (*Mantella aurantica*). Several macrophages have intracytoplasmic vacuoles containing bacteria. Similar bacteria are distributed extracellularly throughout the smear.

lesions also varies and can include classic features of granulomatous inflammation, such as epithelioid macrophages and multinucleated giant cells. In other cases relatively uniform populations of small histiocytic cells that can resemble hematopoietic or lymphoid neoplasia are noted in cytologic or histologic preparations (Fig. 5) [25]. Increased extramedullary hematopoiesis in visceral organs, such as the liver and kidney, can also be observed in affected animals and may complicate cytologic diagnosis. Routine use of stains for acid-fast bacilli (most commonly Ziehl-Neelsen) is recommended for most nodular inflammatory lesions (granulomas and abscesses) and for suspected round cell neoplasms. Rarely, alternative acid-fast stains, such as Fite's, are necessary to demonstrate the organisms commonly implicated in amphibian mycobacteriosis. An inconsistent, but diagnostically helpful, feature of some mycobacterial infections is observation of negative-staining of bacilli in macrophage cytoplasm (Fig. 6).

*Fungal and fungal-like skin diseases*

Skin diseases caused by true fungi or by organisms that are allied with the kingdom Fungi because of morphologic similarity or molecular taxonomy are common in amphibians. As with bacterial skin diseases, fungal infections are often opportunistic and may occur secondary to such factors as cutaneous trauma, suboptimal husbandry, and altered water quality. Most fungal skin diseases are easily amenable to diagnosis by cytologic or wet-mount examination of skin scrapings or aspirates of cutaneous nodules. Careful examination of thicker areas of smears can be helpful in identifying fungal hyphae.

Phaeohyphomycosis and chromomycosis (chromoblastomycosis) are terms used to describe infections with a wide variety of pigmented fungi

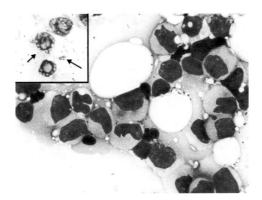

Fig. 5. Impression smear of a vertebral mass in a mountain yellow-legged frog (*Rana muscosa*) showing numerous small histiocytic cells (macrophages). The inset shows a smear stained with a Ziehl-Neelsen acid-fast stain; arrows indicate individual and clumped intracytoplasmic acid-fast bacilli. *Mycobacterium chelonei* was isolated from this lesion.

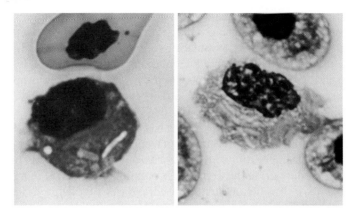

Fig. 6. A monocyte from the peripheral blood of a marine toad (*Bufo marinus*) shows cytoplasmic negative staining that is sometimes observed with mycobacterial infections. An acid-fast stain (*right*) confirms the presence of acid-fast bacilli. *Mycobacterium avium* was isolated from skin lesions in this case. (*Courtesy of* Dr. Leigh Clayton, Baltimore, MD.)

that are saprophytes in soil and plant materials. Chromomycosis in anuran amphibians presents as cutaneous papules or nodules (granulomas) with varying degrees of ulceration [26,27]. Dissemination with granuloma formation in visceral organs can be observed. Microscopically, chromomycosis is characterized by granulomatous inflammation and the presence of round to oval internally septate fungal sclerotic bodies (Fig. 7). Formation of fungal hyphae is reportedly rare with chromomycosis. The author has occasionally recognized phaeohyphomycosis in amphibians as ulcerative skin lesions in debilitated animals. Skin scrapings and histologic sections contained large numbers of light brown septate fungal hyphae without identifiable sclerotic bodies (Fig. 8).

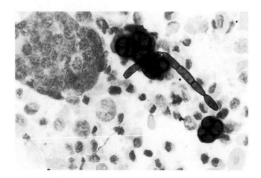

Fig. 7. An impression smear of a raised skin lesion from a spadefoot toad (*Scaphiopus holbrooki*). There are several pigmented, internally septate, spherical sclerotic bodies (suggestive of chromomycosis) and a septate fungal hyphus. A multinucleated giant cell is at the upper left. (*Courtesy of* Dr. Tarja Juopperi, Baltimore, MD.)

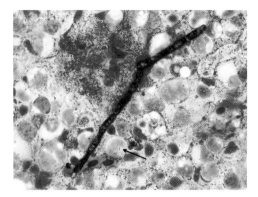

Fig. 8. A mixed bacterial and fungal infection (phaeohyphomycosis) in a forest tree frog (*Leptopelis natalensis*). There is a pigmented septate fungal hyphus, colonies of extracellular bacteria (B), and intracellular bacteria within degenerate macrophages (*arrow*).

Chytridiomycosis is an important emerging disease that has been linked to amphibian population declines on several continents and is caused by the chytrid fungus *Batrachochytrium dendrobatidis* [28]. Histologic lesions of chytridiomycosis consist of epidermal hyperplasia and hyperkeratosis with myriad fungal thalli (bodies) within the cytoplasm of keratinocytes [29]. Because of the superficial location of the infection, preliminary diagnosis of chytridiomycosis can sometimes be accomplished using wet mounts or stained cytologic preparations of skin scrapings or shed skin fragments recovered from the environment [29,30]. Cytologic examination is most useful for heavy, clinically significant infections and is not reliable for diagnosis of low-level subclinical infections. Thalli of *Batrachochytrium* are spherical to flask-shaped and range from 7 to 20 µm in diameter (Figs. 9–11) [31,32].

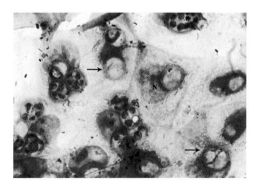

Fig. 9. A sheet of shed skin from a green and black poison dart frog (*Dendrobates auratus*) experimentally infected with the amphibian chytrid fungus *Batrachochytrium dendrobatidis*. There are numerous spherical intracellular chytrid thalli. Many thalli contain small discrete basophilic zoospores and others are empty (*arrows*) having previously discharged zoospores. The empty thallus at the bottom right has evidence of internal septation. N, nucleus of an epithelial cell.

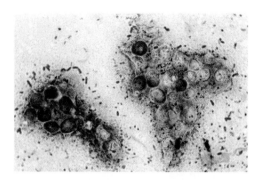

Fig. 10. A skin scraping from a blue poison dart frog (*Dendrobates azureus*) naturally infected with the amphibian chytrid fungus *Batrachochytrium dendrobatidis*. Thalli are both developing and degenerate in this case and may be difficult to distinguish from yeasts. Numerous empty thalli are present on the right side of the photomicrograph.

Mature thalli (zoosporangia) contain numerous 1- to 2-μm diameter discrete basophilic zoospores (see Fig. 9). Empty thalli that have previously discharged zoospores may be observed and occasionally have fine internal septation (see Fig. 9). Developing or degenerate thalli can be difficult to distinguish from yeasts or stages of other organisms including some water molds. Low-level infections are sometimes observed as foci of piled up epithelium (hyperkeratosis), which may be pigmented (see Fig. 11). Use of a Congo red stain has recently been reported to aid in identification of stages of *Batrachochytrium* in skin scrapings from both fresh and formalin-fixed material [33].

Disseminated granulomas associated with the fungus *Mucor amphibiorum* have been reported in anurans from Europe and Australia [34]. The spherical, nonpigmented, approximately 5- to 40-μm, fungal sphaerules frequently

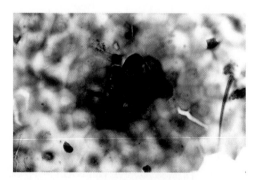

Fig. 11. A sheet of shed skin from a tiger salamander (*Ambystoma tigrinum*) naturally infected with the amphibian chytrid fungus *Batrachochytrium dendrobatidis*. There is a darkly pigmented focus with several retained cell layers (hyperkeratosis) and faintly visible clear spherical structures representing chytrid thalli.

show evidence of endosporulation and differentials include fungi, such as *Coccidioides immitis*, or algae, such as *Prototheca* sp. Hyphae of *M amphibiorum* are not described in amphibian tissues.

Infections with Oomycetes water molds, such as *Saprolegnia, Aphanomyces,* and *Achyla* sp (saprolegniasis), are common in aquatic amphibians and frequently associated with cutaneous injury, low environmental temperatures, high environmental organic loads, and poor water quality [21]. Oomycetes are not true fungi, but rather belong to the kingdom Stramenopila, which also includes the brown algae and diatoms. Infections have a characteristic gross appearance as mats of white to tan cotton-like material that cover cutaneous erosions and ulcers. On wet mounts or in cytologic preparations, Oomycetes hyphae are generally thin-walled, aseptate, rarely branching, and frequently have rounded ends (Fig. 12). Rarely, spherical basophilic zoospores may be observed in association with hyphae. In most cases an associated inflammatory response is minimal or lacking in clinical samples. Because environmental Oomycetes can rapidly colonize organic debris (eg, skin shed into the environment) or dead amphibians, interpretation of cytologic findings should be made in context of sample origin.

Infections with organisms taxonomically situated near the animal-fungal divergence in the class Mesomycetozoea are an important differential for nodular skin or subcutaneous lesions and could be encountered in cytologic preparations of aspirates or impression smears. *Ichthyophonus*-like organisms cause a granulomatous myositis with subsequent body surface swellings [35,36]. On wet mount examination of skeletal muscle squash preparations, organisms are described as cylindrical and thin-walled, measuring an average of 75 × 175 µm [36]. Infections with *Amphibiocystidium ranae* (*Dermocystidium ranae*) are associated with raised cutaneous lesions composed of cysts that contain myriad 2- to 7-µm diameter spherical endospores [37]. Other related organisms, such as *Dermosporidium*, can cause

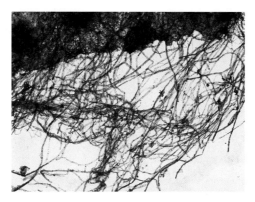

Fig. 12. A sheet of shed skin from a Wyoming toad (*Bufo baxteri*). There is a mat of thin-walled hyphae suggestive of an Oomycete water mold.

similar lesions [38]. Cysts can be associated with varying degrees of granulomatous and granulocytic inflammation.

*Parasitic skin disease*

A wide variety of protozoal and metazoan parasites can be observed in wet mounts and may also appear in stained cytologic preparations of skin scrapings from amphibians [19,21,36]. Protozoa that are more commonly found in diagnostic samples from fish, such as *Trichodina*, *Piscinoodinium*, and *Tetrahymena*, are occasionally encountered in predominantly aquatic amphibians (especially larvae). Infection with the nematode *Pseudocapillaroides xenopi* is well described in the African clawed frog (*Xenopus laevis*) and characteristic bipolar plugged eggs may be found in skin scrapings or on examination of shed skin fragments [39]. Other nematodes, such as *Foleyella* sp or *Dracunculus* sp, and encysted digenetic trematodes can be associated with cutaneous nodules or swellings [19,21]. Common arthropod parasites observed in skin scrapings include trombiculid (chigger) mites and fish lice (*Argulus* sp) [40,41]. Myiasis caused by the larvae of calliphorid flies, such as *Bufolucilia* sp, is also a differential for cutaneous swelling [42].

*Chromatophore hyperplasia*

Changes in cutaneous pigmentation are an occasional clinical presentation in amphibians and are caused by local proliferations (hyperplasia) of one or more of the three types of cutaneous chromatophores (Fig. 13). Melanophores are melanin-producing cells, xanthophores or erythrophores contain carotenoid, and pteridine or flavin pigments and iridophores contain reflective crystalline purines [43,44]. Iridophores can be distinguished

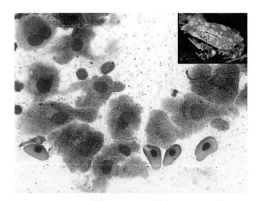

Fig. 13. A skin scraping from an Asian horned frog (*Megophrys nasuta*) with a long history of changes in cutaneous pigmentation manifesting as multifocal raised yellow plaques (*inset*). The scraping shows a population of cells containing a slightly refractile granular cytoplasmic pigment. The tentative diagnosis was chromatophore hyperplasia.

on cytologic examination because they are birefringent under polarized light (Fig. 14). Chromatophore hyperplasia is usually a nonspecific finding and can occur at sites of previous cutaneous injury or infection or in association with cutaneous parasitism. In some cases the cause cannot be determined and the differential diagnosis includes pigment cell neoplasia.

## Subcutaneous lymph sac and coelomic effusions

Fluid accumulations in the subcutaneous lymph sacs or coelomic cavity are among the more common clinical presentations in captive amphibians. These effusions are associated with an extensive differential diagnosis that can include bacterial septicemia, heart failure, lymph heart failure, renal or hepatic disease, hypocalcemia, hypoproteinemia, and osmotic imbalances associated with housing in low solute water [45].

Comprehensive fluid analysis that includes a cell count, determination of total protein, and cytologic examination can provide essential information for clinical management of these cases. For effusions of low cellularity (usually clear and colorless) examination of centrifuged sediment smears is ideal. Experience is limited with clinicopathologic correlation in regard to amphibian fluid samples and reference values have not been established; however, use of guidelines established for domestic animals can (cautiously) be an approximate starting point for interpretation of fluid analyses [46]. In general, samples with a very low total protein (<2.5 g/dL) and low cellularity (<1500 cells/µL) can be classified as transudates and are most consistent with hypoproteinemia, biochemical, and physiologic causes of effusion and renal disease, among others. Samples with higher total protein (>3 g/dL) and cell counts (> 7000/µL) are probably exudates and inflammatory conditions should be considered. Examination of stained smears for degenerate inflammatory cells and intracellular bacteria can be helpful when

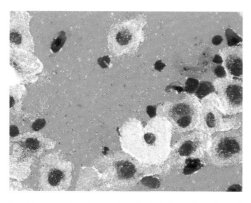

Fig. 14. Cells from the skin scraping shown in Fig. 13 viewed under polarized light. Pigment granules are birefringent indicating that cells are iridophores.

considering bacterial septicemia as a potential cause for effusion. As for domestic animals, low numbers of mesothelial cells (for coelomic effusions), macrophages, granulocytes, and lymphocytes may be observed even in noninflammatory effusions (see Fig. 1).

## Fecal and cloacal wash cytology

Examination of stained smears of feces, cloacal washes, or intestinal scrapings are useful primarily as an aid in the diagnosis of some forms of enteric parasitism, such as coccidiosis [19]. Cryptosporidiosis has only rarely been documented in amphibians; however, presumably a modified acid-fast stain on fecal smears may be helpful in demonstrating oocysts as described for other species [47,48]. The author has found wet mount and cytologic examination of distal intestinal contents useful for rapid diagnosis of strongyloidiasis in recently imported anurans (Fig. 15) [49]. Observation of leukocytes and erythrocytes in fecal smears can suggest underlying intestinal tract inflammation; however, the absence of observed cells does not exclude the presence of a significant inflammatory lesion. Many ciliate and flagellate protozoa can be normal commensal organisms in amphibian intestinal tracts [50].

## Cytology of visceral organs

Although much of amphibian diagnostic cytology is focused on the integumentary system, fine-needle aspirates and impression smears of lesions from visceral organs and cytologic examination of samples collected in procedures, such as tracheal washes, are becoming increasingly common as amphibian medicine becomes more sophisticated. In addition, use of visceral

Fig. 15. A smear of intestinal contents from a milky tree frog (*Phrynohyas resinifictrix*) showing a nematode parasite later identified as a *Strongyloides* sp.

organ impression smears collected at necropsy can be invaluable in providing a rapid diagnosis while awaiting results of histopathology or ancillary diagnostic tests (Fig. 16). A brief review of some normal features of amphibian visceral organ cytology and some common or emerging diseases that may be diagnosed by cytologic examination are presented.

## Normal cytologic features of visceral organs

A few normal features of amphibian viscera may be confusing when encountered on impression smears and are worthy of mention. Melanomacrophages are unique melanin-producing cells of the mononuclear phagocyte system found in fish, amphibians, and reptiles [51–53]. In amphibians, they are most often observed as discrete aggregates (melanomacrophage centers) in the liver and to a lesser degree the spleen; however, it is not unusual to observe small numbers of pigmented macrophages in a variety of tissues (see Fig. 1). Increases in the size of melanomacrophage centers (melanomacrophage hyperplasia) are common and nonspecific and may be associated with seasonal variation, emaciation, or antigenic stimulation. In cytologic preparations, melanomacrophages frequently rupture and free brown-black pigment granules may be observed in the background of smears (Fig. 17). Extramedullary hematopoiesis of both granulocytic and erythrocytic cell lines is common in the liver and kidney and samples that include foci of granulopoiesis can be misdiagnosed as inflammatory lesions. Extramedullary hematopoiesis can be exceptionally prominent in juvenile animals and animals with persistent antigenic stimulation. Both glycogen and lipid-type vacuolar changes are common in amphibian hepatocytes (especially larvae) and generally should not be interpreted as having clinical significance (see Fig. 2).

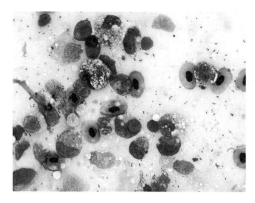

Fig. 16. An impression smear of the spleen in a Panamanian golden frog (*Atelopus zeteki*) with splenomegaly. Several macrophages contain intracytoplasmic bacteria suggestive of septicemia.

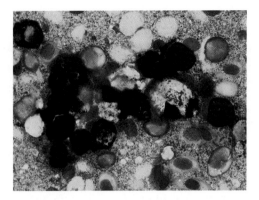

Fig. 17. An impression smear of the liver from a mountain-yellow legged frog (*Rana muscosa*) showing several melanomacrophages and extracellular black pigment from artifactually disrupted melanomacrophages. An erythroid precursor cell is at the upper right corner and suggests extramedullary hematopoiesis.

## Bacterial and fungal diseases

Systemic bacterial infections are common in debilitated, stressed, or crowded amphibians and postmortem impression smears of liver and spleen can occasionally provide rapid confirmation of this diagnosis by demonstration of intracellular bacteria in macrophages or granulocytes (see Fig. 16). Mycobacteriosis and some fungal infections, such as chromomycosis and mucormycosis, frequently are characterized by granulomas in multiple visceral organs and are amenable to diagnosis by cytology.

## Chlamydophilosis

Systemic infections with *Chlamydophila psittaci* and *Chlamydophila pneumoniae* have been described in African clawed frogs (*Xenopus* sp) and occasionally other anuran species [54–56]. Most diagnoses have been initiated by histopathologic examination of formalin-fixed tissues with observation of granular basophilic cytoplasmic material (inclusions) in hepatocytes; splenic reticuloendothelial cells; or macrophages. It is likely that inclusions suggestive of *Chlamydophila* infection could be encountered in organ impression smears of infected frogs as has been observed in other animal groups [57,58].

## Protozoal and myxozoal diseases

An organism described as "*Perkinsus*-like" on the basis of 18s ribosomal RNA sequencing has been reported as the cause of mass mortality events in tadpoles of ranid frogs in the United States [15,59]. Affected animals have hepatosplenomegaly and renomegaly with massive infiltrates of spherical basophilic organisms (Fig. 18). There are minimal or no associated

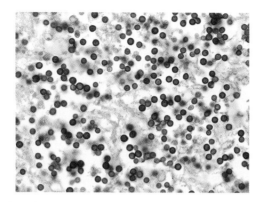

Fig. 18. A histologic section of the liver from a Mississippi gopher frog (*Rana sevosa*) tadpole showing numerous spherical basophilic *Perkinsus*-like protozoal organisms. This organism can also be identified in organ impression smears.

inflammatory cell infiltrates. Organisms can be observed on cytology of organ impression smears or histopathology of affected animals.

Myxozoa are microscopic metazoan parasites thought to be related to the cnidarians. Although often observed as incidental findings, Myxozoa in the genera *Chloromyxum* and *Hoferellus* have recently been described as causes of tubulointerstitial nephritis [60,61]. Spores with characteristic polar capsules and occasional extrusion of polar filaments can be observed on impression smears of the kidney. Other potentially important Myxozoa of amphibians occur in the reproductive tract or are associated with the bile ducts and gallbladder [19,62].

*Nematode infections*

Pulmonary infections with the rhabditiform nematode *Rhabdias* sp are common in captive amphibians and heavy parasite loads can be clinically significant [63]. Larvated eggs, larvae, and adult worms may be recovered on tracheal washes of larger animals.

## Noninfectious diseases

Approaches toward cytologic diagnosis of noninfectious diseases, most often neoplasia, by cytologic examination are similar to those used in other species. A wide variety of neoplastic diseases have been described in amphibians and these have recently been reviewed [25]. In most instances, neoplasia is a differential along with fungal and bacterial diseases for nodular masses either in the skin or visceral organs. In general, neoplasms can be tentatively classified on the basis of cell shape and cell-to-cell interactions as epithelial (cells arranged in sheets or clumps); spindle cell (individualized cells, elongate cell shape); or round cell (individualized cells, round cell shape).

Individual cell criteria for malignancy can be used with caution recognizing that the biologic behavior and appearance of amphibian neoplasms is not always well documented. Histiocytic cells in some cases of mycobacteriosis can resemble hematopoietic neoplasia and caution should be used in interpretation.

## References

[1] Wright KM. Amphibian hematology. In: Wright KM, Whitaker BR, editors. Amphibian medicine and captive husbandry. Malabar (FL): Krieger Publishing; 2001. p. 129–46.
[2] Jenkins J. Basic amphibian hematology: an introduction to techniques of collection, preparation and identification. Proc Assoc Zoo Vet Tech 1997;13–9.
[3] Turner RJ. Amphibians. In: Rowley AF, Ratcliffe NA, editors. Vertebrate blood cells. Cambridge: Cambridge University Press; 1988. p. 129–209.
[4] Wojtaszek J, Adamowicz A. Hematology of the fire-bellied toad, *Bombina bombina* L. Comp Clin Pathol 2003;12:129–34.
[5] Montali RJ. Comparative pathology of inflammation in the higher vertebrates (reptiles, birds and mammals). J Comp Pathol 1988;99:1–26.
[6] Campbell TW. Clinical pathology of reptiles. In: Mader DR, editor. Reptile medicine and surgery. St Louis: Saunders; 2006. p. 453–70.
[7] Desser SS. *Aegyptianella ranarum* sp. n. (Rickettsiales, Anaplasmataceae): ultrastructure and prevalence in frogs from Ontario. J Wildl Dis 1987;23:52–9.
[8] Desser SS, Barta JR. The morphological features of *Aegyptianella bacterifera*: an erythrocytic rickettsia of frogs from Corsica. J Wildl Dis 1989;25:313–8.
[9] Babudieri B. *Mycoplasma*-like organism, parasite of red blood cells of an amphibian, *Hydromantes italicus*. Infect Immun 1972;6:77–82.
[10] Zhang C, Rikihisa Y. Proposal to transfer *Aegyptianella ranarum*, an intracellular bacterium of frog red blood cells to the family Flavobaceriaceae as Candidatus Hemobacterium ranarum comb. nov. Environ Microbiol 2004;6:568–73.
[11] Speare R, Freeland RJ, Bolton SJ. A possible iridovirus in erythrocytes of *Bufo marinus* in Costa Rica. J Wildl Dis 1991;27:457–62.
[12] Gruia-Gray J, Desser SS. Cytopathological observations and epizootiology of frog erythrocytic virus in bullfrogs (*Rana catesbeiana*). J Wildl Dis 1992;28:34–41.
[13] Telford SR, Jacobson ER. Lizard erythrocytic virus in east African chameleons. J Wildl Dis 1993;29:57–63.
[14] Docherty DE, Meteyer CU, Wang J, et al. Diagnostic and molecular evaluation of three iridovirus-associated salamander mortality events. J Wildl Dis 2003;39:556–66.
[15] Green DE, Converse KA, Schrader AK. Epizootiology of sixty-four amphibian morbidity and mortality events in the USA, 1996–2001. Ann N Y Acad Sci 2002;969:323–39.
[16] Allender MC, Fry MM, Irizarry AR, et al. Intracytoplasmic inclusions in circulating leukocytes from an eastern box turtle (*Terrapene carolina carolina*) with iridoviral infection. J Wildl Dis 2006;42:677–84.
[17] Desser SS. Ultrastructural observations on a icosahedral cytoplasmic virus in leukocytes of frogs from Algonquin Park, Ontario. Can J Zool 1992;70:833–6.
[18] Desser SS. The blood parasites of anurans from Costa Rica with reflections on the taxonomy of their trypanosomes. J Parasitol 2001;87:152–60.
[19] Poynton SL, Whitaker BR. Protozoa and metazoa infecting amphibians. In: Wright KM, Whitaker BR, editors. Amphibian medicine and captive husbandry. Malabar (FL): Krieger Publishing; 2001. p. 193–221.
[20] Smith TG. The genus *Hepatozoon* (Apicomplexa: Adeleina). J Parasitol 1996;82:565–85.

[21] Pessier AP. An overview of amphibian skin disease. Semin Avian Exot Pet Med 2002;11: 162–74.
[22] Reavill DR. Amphibian skin diseases. Vet Clin Exot Anim 2001;4:413–40.
[23] Whitaker BR, Wright KM. Clinical techniques. In: Wright KM, Whitaker BR, editors. Amphibian medicine and captive husbandry. Malabar (FL): Krieger Publishing; 2001. p. 89–110.
[24] Ferreira R, Fonseca LD, Afonso AM, et al. A report of mycobacteriosis caused by *Mycobacterium marinum* in bullfrogs (*Rana catesbeiana*). Vet J 2006;171:177–80.
[25] Stacy BA, Parker JM. Amphibian oncology. Vet Clin Exot Anim 2004;7:673–95.
[26] Miller EA, Montali RJ, Ramsay EC, et al. Disseminated chromoblastomycosis in a colony of ornate-horned frogs (*Ceratophrys ornata*). J Zoo Wildl Med 1992;23:433–8.
[27] Juopperi T, Karli K, DeVoe R, et al. Granulomatous dermatitis in a spadefoot toad (*Scaphiopus holbrooki*) [chromomycosis]. Vet Clin Pathol 2002;31:137–9.
[28] Lips KR, Brem F, Brenes R, et al. Emerging infectious disease and the loss of biodiversity in a neotropical amphibian community. Proc Natl Acad Sci U S A 2006;103:3165–70.
[29] Pessier AP, Nichols DK, Longcore JE, et al. Cutaneous chytridiomycosis in poison dart frogs (*Dendrobates* sp.) and White's tree frogs (*Litoria caerulea*). J Vet Diagn Invest 1999; 11:194–9.
[30] Nichols DK, Lamirande EW, Pessier AP, et al. Experimental transmission of cutaneous chytridiomycosis in dendrobatid frogs. J Wildl Dis 2001;37:1–11.
[31] Berger L, Hyatt AD, Speare R, et al. Life cycle stages of the amphibian chytrid *Batrachochytrium dendrobatidis*. Dis Aquat Organ 2005;68:51–63.
[32] Davidson EW, Parris M, Collins JP, et al. Pathogenicity and transmission of chytridiomycosis in tiger salamanders (*Ambystoma tigrinum*). Copeia 2003;2003:601–7.
[33] Briggs C, Burgin S. Congo red, an effective stain for revealing the chytrid fungus, *Batrachochytrium dendrobatidis*, in epidermal skin scrapings from frogs. Mycologist 2004;18: 98–103.
[34] Speare R, Thomas AD, O'Shea P, et al. *Mucor amphibiorum* in the toad *Bufo marinus* in Australia. J Wildl Dis 1994;30:399–407.
[35] Mikaelian I, Ouellet M, Pauli B, et al. *Ichthyophonus*-like infection in wild amphibians from Quebec, Canada. Dis Aquat Organ 2000;40:195–201.
[36] Green DE. Pathology of amphibia. In: Wright KM, Whitaker BR, editors. Amphibian medicine and captive husbandry. Malabar (FL): Krieger Publishing; 2001. p. 401–85.
[37] Pereira CN, Di Rosa I, Fagotti A, et al. The pathogen of frogs *Amphibiocystidium ranae* is a member of the order Dermocystida in the class Mesomycetozoea. J Clin Microbiol 2005; 43:192–8.
[38] Jay JM, Pohley WJ. *Dermosporidium penneri* sp. n. from the skin of the American toad, *Bufo americanus* (Amphibia: Bufonidae). J Parasitol 1981;67:108–10.
[39] Stephens LC, Cromeens DM, Robbins VW, et al. Epidermal capillariasis in South African clawed frogs (*Xenopus laevis*). Lab Anim Sci 1987;37:341–4.
[40] Sladky KK, Norton TM, Loomis MR. Trombiculid mites (*Hannemania* sp.) in canyon tree frogs (*Hyla arenicolor*). J Zoo Wildl Med 2000;31:570–5.
[41] Wolfe BA, Harms CA, Groves JD. Treatment of *Argulus* sp. infestation of river frogs. Contemp Top Lab Anim Sci 2001;40:35–6.
[42] Bolek MG, Coggins JR. Observations on myiasis by the calliphorid, *Bufolucilia silvarum*, in the Eastern American toad (*Bufo americanus americanus*) from Southeastern Wisconsin. J Wildl Dis 2002;38:598–603.
[43] Hoffman EA, Blouin MS. A review of colour and pattern polymorphisms in anurans. Biological Journal of the Linnean Society 2000;70:633–65.
[44] Irizarry-Rovira AR, Wolf A, Ramos-Vara JA. Cutaneous melanophoroma in a green iguana (*Iguana iguana*). Vet Clin Pathol 2006;35:101–5.
[45] Wright KM. Idiopathic syndromes. In: Wright KM, Whitaker BR, editors. Amphibian medicine and captive husbandry. Malabar (FL): Krieger Publishing; 2001. p. 239–44.

[46] Cowell RL, Tyler RD, Meinkoth JH. Abdominal and thoracic fluid. In: Cowell RL, Tyler RD, Meinkoth JH, editors. Diagnostic cytology and hematology of the dog and cat. 2nd edition. St. Louis: Mosby; 1999. p. 142–58.

[47] Green SL, Bouley DM, Josling CA, et al. Cryptosporidiosis associated with emaciation and proliferative gastritis in a laboratory-reared South African clawed frog (*Xenopus laevis*). Comp Med 2003;53:81–4.

[48] Harr KE, Henson KL, Raskin RE, et al. Gastric lavage from a Madagascar tree boa (*Sanzinia madagascarensis*) [cryptosporidiosis]. Vet Clin Pathol 2000;29:93–6.

[49] Patterson-Kane JC, Eckerlin RP, Lyons ET, et al. Strongyloidiasis in a Cope's grey tree frog (*Hyla chrysoscelis*). J Zoo Wildl Med 2001;32:106–10.

[50] Poynton SL, Whitaker BR. Protozoa in poison dart frogs (Dendrobatidae): clinical assessment and identification. J Zoo Wildl Med 1994;25:29–39.

[51] Agius C, Roberts RJ. Melano-macrophage centers and their role in fish pathology. J Fish Dis 2003;26:499–509.

[52] Barni S, Vaccarone R, Bertone V, et al. Mechanisms of changes to the liver pigmentary component during the annual cycle (activity and hibernation) of *Rana esculenta* L. J Anat 2002;200:185–94.

[53] Gyimesi ZS, Howerth EW. Severe melanomacrophage hyperplasia in a crocodile lizard, *Shinisaurus crocodilurus*: a review of melanomacrophages in ectotherms. J Herp Med Surg 2004;14:19–23.

[54] Howerth EW. Pathology of naturally-occurring chlamydiosis in African clawed frogs (*Xenopus laevis*). Vet Pathol 1984;21:28–32.

[55] Berger L, Volp K, Mathews S. *Chlamydia pneumoniae* in a free-ranging giant barred frog (*Mixophyes iteratus*) from Australia. J Clin Microbiol 1999;37:2378–80.

[56] Bodetti TJ, Jacobson ER, Wan C, et al. Molecular evidence to support the expansion of the host range of *Chlamydophila pneumoniae* to include reptiles as well as humans, horses, koalas and amphibians. Syst Appl Microbiol 2002;25:146–52.

[57] Strik NI, Alleman AR, Wellehan JFX. Conjunctival swab cytology from a guinea pig: it's elementary! Vet Clin Pathol 2005;34:169–71.

[58] Campbell TW. Cytology of internal organs and lymphoid tissue. In: Avian hematology and cytology. 2nd edition. Ames (IA): Iowa State University Press; 1995. p. 74–80.

[59] Green DE, Feldman SH, Wimsatt J. Emergence of a *Perkinsus*-like agent in anuran liver during die-offs of local populations: PCR detection and phylogenetic characterization. Proc Am Assoc Zoo Vet 2003;120–1.

[60] Duncan AE, Garner MM, Bartholomew JL, et al. Renal myxosporidiasis in Asian horned frogs (*Megophrys nasuta*). J Zoo Wildl Med 2004;35:381–6.

[61] Mutschmann F. Pathological changes in African hyperoliid frogs due to a myxosporidian infection with a new species of *Hoferellus* (Myxozoa). Dis Aquat Organ 2004;60:215–22.

[62] Hill BD, Green PE, Lucke HA. Hepatitis in the green tree frog (*Litoria caerulea*) associated with infection by a species of *Myxidium*. Aust Vet J 1997;75:910–1.

[63] Nichols DK. Amphibian respiratory diseases. Vet Clin Exot Anim 2000;3:551–4.

# Diagnostic Cytology of Fish

Drury Reavill, DVM, DABVP, DACVP[a],*, Helen Roberts, DVM[b]

[a]Zoo/Exotic Pathology Service, 7647 Wachtel Way, Citrus Heights, CA 95610, USA
[b]5 Corners Animal Hospital, 2799 Southwestern Boulevard, Suite 100, Orchard Park, NY 14127, USA

Cytology is the cornerstone of diagnostic techniques in fish medicine. It involves the evaluation of patient cells and an assessment of associated bacterial and fungal populations, protozoan infestations, and metazoan parasites or their eggs. The benefits of cytology include easy sample collection with minimal patient risk, relatively low cost, and rapid results. Interpretation of findings can help to establish a diagnosis and initiate a rapid therapeutic plan.

## Methods and indications

The most common cytologic preparation is of samples from the gill(s) and skin of the fish for parasite identification. Other indications include quick evaluation of masses, cutaneous lesions or ulcers, and fluid accumulations in the coelomic cavity or organs. When performing a skin scrape, always handle the fish with gloves. This reduces the loss of protective mucus coating from the fish's skin and also protects the clinician. For large fish, anesthesia may be required to provide adequate restraint, prevent the practitioner from dropping the fish, and protect the fish from self-inflicted injuries.

### Wet mount preparation

Place a drop of water from the fish's tank or pond on a clean slide. Tap water may contain chlorine or other additives that can destroy the parasites. Use a coverslip or the blunt side of a scalpel blade to scrape the skin gently

* Corresponding author.
  *E-mail address:* dreavill@zooexotic.com (D. Reavill).

in a cranial-caudal direction. It has been suggested that most parasites accumulate in areas subject to less drag in the water. Areas that may yield a high number of ectoparasites include skin behind the fins, on the caudal peduncle or tail, and under the chin. Place the collected sample on the slide underlying a coverslip, and microscopically examine it with original magnification ×4 to ×100 objectives (Fig. 1).

## Gill snip

Using suture scissors or iris tenotomy scissors, remove the tips of the gill lamellae and one layer only. This should be of a sufficient size to see primary and secondary lamellae. Alternatively, use a coverslip to scrape the gills gently to acquire the sample. Be sure to warn the client that some bleeding may occur from the scrape or clip, particularly in koi (*Cyprinus carpio*) (Helen Roberts, DVM, Orchard Park, NY, personal communication, 2006). Place the sample on a clean slide, and cover it with a coverslip. Microscopically, examine the gill tissue with the original magnification ×4 to ×100 objectives (Fig. 2).

Other cytology collection methods are similar to those used with domestic animal species. These include the isolation of a mass or structure to be aspirated, collection with an appropriately sized needle and syringe, and placement onto a glass slide for direct or stained examination. If excess mucus or a slime layer is present on the lesion, removal can be accomplished by gently wiping the area with a sterile gauze sponge. Scraping the cut surface of more firm masses may be necessary to obtain cells for examination. Impression smears are useful, especially in postmortem samples. There is an additional classic form of cytology preparation that is valuable for rapid

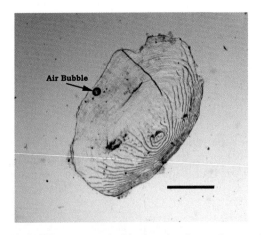

Fig. 1. Fish scale. Air bubbles are round with sharp borders and are variably shaped. Wet mount. Bar = 0.5 mm.

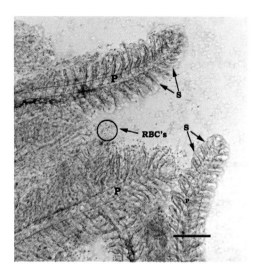

Fig. 2. Gill. The section has the primary lamella (P) and secondary lamella (S). Red blood cells (RBCs) are streaming out from the cut sections. Wet mount. Bar = 0.1 mm.

diagnosis at postmortem examination of fish. This is the tissue squash method. Small cut samples of the liver, intestines, or kidney can be placed on a slide and gently squashed with a coverslip (Figs. 3 and 4). These are examined as wet mount preparations or can be stained. Parasites and granulomas are the most common findings with this method (Fig. 5).

A cytology preparation should be collected from masses or structures (eg, cysts, gonads, eyes) that are submitted for histologic examination. This is an excellent method of continuing education for the practitioner and the pathologist in the cytology of fish diseases. The most common conditions are covered in this article. There are excellent references for more in-depth

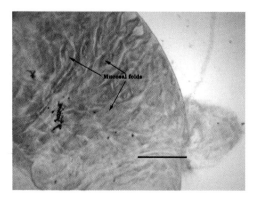

Fig. 3. Intestine from a tetra. The mucosal folds fill the lumen. Wet mount. Bar = 0.5 mm.

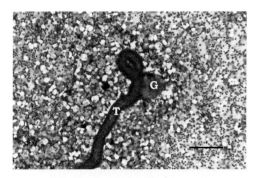

Fig. 4. Squash preparation of the hematopoietic kidney of a goldfish. G, glomerulus; T, tubule. The dense cellular background represents the hematopoietic elements. May-Grunwald Giemsa stain. Bar = 0.1 mm.

evaluation of infectious disease agents [1,2], neoplastic lesions [3], and diagnostic techniques [4].

*Coelomic cavity*

When fluid accumulates within the coelomic cavity of fish, it can be aspirated and examined cytologically. The most common cause of ascitic fluid accumulation is from systemic bacterial infections. In affected fish, the cytologic findings are of mixed inflammatory cell populations, with or without microorganisms (Fig. 6). Hemorrhagic fluid may be observed with some tumors of the coelomic cavity.

In electric eels (*Electrophorus electricus*), coelomic fluid accumulation has resulted from increased body sodium because of increased water pH. In these freshwater eels, the hydrogen ions may act as counter-ions to sodium uptake. The increased body sodium results in fluid retention as well as cardiac overload, with the development of coelomic cavity ascites. The coelomic fluid has a specific gravity of 1.013 to 1.016 [5].

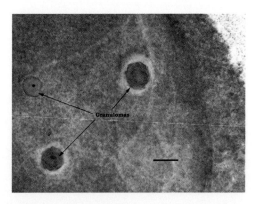

Fig. 5. Liver from a tetra. Three bacterial granulomas are present. Wet mount. Bar = 0.1 mm.

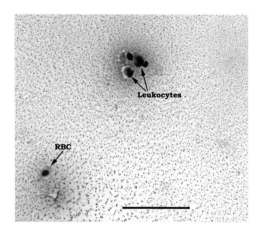

Fig. 6. Ascitic fluid attributable to renal failure. A few cells are recognized on a background of proteinaceous fluid. RBC, red blood cell. May-Grunwald Giemsa stain. Bar = 0.05 mm.

*Digestive tract*

The digestive tract is similar to that of mammals, with the length depending on diet type (herbivores, carnivores, or omnivores). Herbivorous fish tend to have longer guts than carnivores. In some species, there are extensions of the pyloric portion of the stomach called pyloric cecae. Cyprinids, such as koi and goldfish, do not have a true stomach. Parasites, infections from bacteria or fungus, and tumors may be identified within the digestive tract.

Evaluation of the feces can be performed as described for other species to identify internal parasites. Many fish defecate when anesthetized. If no stool is available, a cloacal wash can be evaluated. At postmortem examination, an intestinal tissue squash preparation can be used to identify parasites and bacterial or fungal infections. This technique is useful for rapid diagnostics in an outbreak of disease within a collection of fish.

Several metazoan parasites can be found in the digestive tract of ornamental fish. Adult and larval nematodes can be found within the lumen of the intestinal tract, migrating through the coelomic cavity or encysted in internal organs or musculature. Some examples include *Eustrongyloides* species, which are frequently found encysted in the muscles of the peritoneum and are considered an incidental finding; *Capillaria* species, which are found frequently in the intestinal lumen of ornamentals but is of unknown significance (Fig. 7) [6]; and *Camallanus* species, the blood red worm frequently seen protruding from the vent of live-bearing fish, such as guppies (*Poecilia reticulata*) and swordtails (*Xiphophorus helleri*).

*Camallanus* species are probably the best recognized nematodes of freshwater tropical fish [7]. They are not host specific and have been described in elasmobranchs, such as an aquarium-reared stingray *Potamotrygon* sp

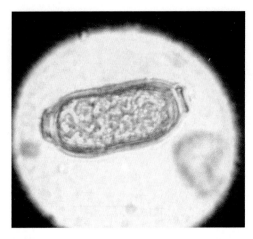

Fig. 7. Fecal examination. Capillaria egg with the plug at each end. Wet mount.

(Dasyatididae) [8]. The female nematodes are viviparous, producing large embryos that are released as live young [1,2]. The larva are ingested by an intermediate host, often a crustacean, and then by a mature fish (Fig. 8).

Encysted parasites (trematodes, cestodes, and nematodes) can be identified on an intestinal squash preparation. The metazoan parasites are generally degenerate depending on the length of time they have been encysted and the host response. A variably thick capsule of inflammatory cells and fibrous connective tissue might be recognized.

Protozoa are a prominent part of the fish digestive tract microflora. Several are known pathogens but most are considered normal flora. Cryptosporidiosis has been described in several freshwater fish, including *Plecostomus* species [9,10]. Weight loss and emaciation are common clinical signs. *Cryptosporidia* infections have also been recognized in sea horses (*Hippocampus capensis*) [11] and other marine fish [12]. Although not reported, it is

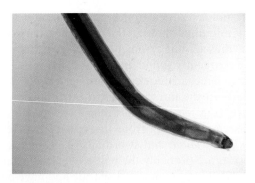

Fig. 8. Camallanus are bright red. Wet mount.

expected that a fecal examination may result in identification of *Cryptosporidia* spores. An acid-fast stain of a fecal smear should be considered for fish that are losing weight.

Coccidial infections are relatively common in goldfish [13] and are also recognized in freshwater and saltwater species [14,15]. Infections can be severe depending on the age of the fish, nutritional status, and environmental stressors. Coccidia primarily have a direct life cycle; however, a few have indirect life cycles with the intermediate host, palaemonid shrimp [16]. The direct life cycle is complete by ingestion of the sporulated oocysts. Loss of body condition, particularly muscle atrophy, is a common finding. A fecal examination or intestinal or cloacal wash may be helpful in picking up these organisms, because they are shed through the digestive tract.

*Spironucleus* (Hexamita) is a flagellated protozoan found in the intestinal tract and gallbladder of a variety of fish [17–20]. In many fish, such as trout (*Oncorhynchus* species), goldfish (*Carassius auratus*), and a variety of freshwater ornamentals, the infestations are inapparent. In other ornamentals, including angelfish (*Pterophyllum* species), discus fish (*Symphysodon* species), and gouramis (*Colisa* species), it causes poor body condition, inappetence, unthriftiness, weight loss, and death. These clinical signs are attributable to the associated enteritis and, occasionally, systemic invasion. In discus fish, the intestinal proliferations are associated with hole-in-the-head disease after a systemic infection [18]. The parasites are the size of red blood cells and have rapid forward hypermotility. They have a pyriform shape with eight flagella (difficult to identify on wet mounts).

There are few reports of gastrointestinal tumors in fish. Nevertheless, masses should be aspirated and examined cytologically to determine if they may be attributable to inflammation from infectious disease agents (bacteria, fungus, or metazoans) or a neoplastic lesion. This has some bearing on the prognosis and importance to other fish in a pond or aquarium system [21,22].

*Integument*

The skin and fins are the most common sites for exfoliative cytology collection. The samples collected should include wet mounts, stained cytologic samples, and cultures (bacterial or fungal). For masses that can be easily removed, cytology and concurrent histopathology help with diagnosis and therapy. Some conditions that can be diagnosed include those caused by disease agents (virus, bacteria, fungus, or parasites) and tumors.

Herpesvirus-associated papillomas in koi (also known as carp pox) are induced by the cyprinid herpesvirus 1 (CyHV-1) [23]. This is a lesion of epidermal hyperplasia. The incubation period is typically 2 to 10 months. In adult fish, there is little morbidity or mortality. Younger fish can experience anywhere from a 20% to 97% fatality rate. Transformation to squamous cell carcinoma is uncommonly reported. These proliferative cutaneous lesions range from single to multiple, white to pink, fleshy masses that are typically associated with the caudal dorsal pectoral, anal fins, and head. They may spontaneously resolve within 1 to 3 weeks, although some can persist for longer than 3 months. On cytology preparations, the lesions are densely cellular and consist of a hyperplastic epidermal cell with increased cytoplasmic basophilia. Rarely, parasites or inflammatory cells are identified; however, these are typically secondary processes. Histologically, these biopsies have marked benign epidermal hyperplasia without significant inflammatory infiltrates. Viral inclusion bodies are not typically identified [24,25]. Koi herpesvirus (KHV; CyHV-3) also has skin lesions that appear as pale, raised, rough patches of scaleless skin [24]. A description of the cytologic and histologic findings is not published. Other lesions are described in the section on the respiratory system.

Lymphocystis is caused by an iridovirus and is a chronic and generally benign lesion of the fibroblasts. It can infect fresh and marine fish [26]. These lesions are usually cutaneous but have also been identified within the muscle, peritoneum, or visceral organs [27–29]. Grossly, the cutaneous masses are pearlescent cream to pink or tan nodules. On cytology, there is marked hypertrophy of dermal connective tissue cells. These are characterized as round cells of various sizes, ranging from 100 to 500 µm in diameter, with a smooth capsule. The surrounding tissues may support inflammatory infiltrates of macrophages or lymphocytes as well as plasma cells. These clusters of large cells may be difficult to differentiate from epitheliocystis, a lesion on the gills and, occasionally, on the skin, which is presumably caused by a *Chlamydia*-like microbe. Histologically, these can be differentiated; lymphocystis cells are oval to spherical and surrounded by a thick laminated periodic acid–Schiff (PAS)-positive hyaline capsule (Fig. 9). The cell nucleus and nucleolus remain centrally located but are also markedly hypertrophied. Viral inclusion bodies can be seen on histology and are typically cytoplasmic cord-like structures. This virus is transmitted by direct contact and is facilitated by skin trauma. It is a self-limiting viral infection unless the lesion involves the oral cavity and restricts food intake. The recommendation is to isolate the fish for approximately 6 weeks and to remove the growths surgically, monitoring for secondary bacterial infections.

Bacterial infections are major contributors to mortality in aquarium and pond fish. The most common disease-causing bacteria belong to the gram-negative aerobes or facultative anaerobes. These bacteria also exist ubiquitously in the fish's environment; they are natural saprophytes in the water. Many bacteria cultured from external lesions are opportunistic pathogens, and it is not uncommon to culture coliforms. Generally, any stressors, including parasitic infections, malnutrition, recent transportation, temperature extremes, social stress, overcrowding, poor water quality, and trauma, can result in a bacterial disease. The clinical signs and the lesions of bacterial infections are generally similar for all the common isolates. Most of these result in a septicemic or ulcerative disease process.

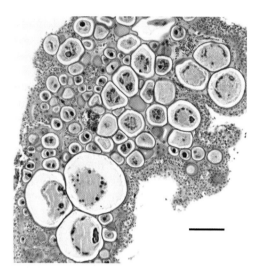

Fig. 9. The cells of lymphocystis are oval to spherical and surrounded by a thick laminated capsule. Hematoxylin-eosin section from a skin mass. Bar = 0.1 mm.

Hemorrhages, skin ulcers, fin and tail rot, ascites, exophthalmia, and a change in normal color are commonly seen. The fish is often listless and anorectic. Treatment is based on the culture and sensitivity test results and on resolving underlying issues (husbandry and poor water quality) [30].

The *Aeromonas* complex (bacterial hemorrhagic septicemia and ulcer disease) is a group of motile, facultative, anaerobic, gram-negative bacterial rods that are ubiquitous organisms affecting all freshwater fish [31]. Some strains are also halophilic tolerant and occasionally cause disease in saltwater fish. The aeromonads may be the most common bacterial pathogens in captive fish. *Aeromonas hydrophilia* is the most common isolate of the group that also includes atypical *Aeromonas salmonicida* and *Aeromonas sobria* [32]. Several other species are described within the motile *Aeromonas* complex, but complete classification is under investigation [33,34]. Disease attributable to *Aeromonas* species occurs most often as a secondary invader in association with some type of environmental stress or injury to the fish, and in waters with a high organic content [35]. The disease ranges from peracute to chronic presentations. It can also involve a single animal or run an epidemic throughout an entire fish community. Mortality and morbidity rates can be high. The signs typify those of any bacterial infection: exophthalmia, skin ulcers, ascites, and hemorrhages on the body surface. The exfoliative cytology preparations of external lesions have inflammatory cells, bacteria, and possibly protozoans.

*Flavobacterium columnare* (formerly known as *Flexibacter columnaris*) is a highly infectious bacterial infection of tropical fish. These are gram-negative filamentous bacteria that exhibit flexing movement. The fish develops a grayish-white discoloration, especially on the head and around the mouth, earning it the common name of "cotton wool disease." Common stressors associated with an infection are temperature extremes, stress from shipping and high stocking density, elevated water ammonia levels, and an increased organic load. This bacterium is transmitted by direct contact. It can lead to a septicemic process when it enters the bloodstream, particularly through gill lesions. Adhesion to gill tissue is enhanced by elevated nitrite levels in the water. Typically, the highly virulent strains result in extensive gill necrosis. On cytology preparations of affected skin, gills, and fin, the lesions support a dense population of long, slender, filamentous, rod-shaped bacteria that aggregate in columns or mounds resulting in haystacks. A variable inflammatory cell population may be present. Culture isolation is difficult, because this particular bacterium requires special media. The therapy involves removing stressful conditions, improving water quality, and the administration of antibacterials based on drug sensitivies [2].

*Mycobacterium* species also result in significant skin lesions as well as systemic lesions. These bacteria are nonmotile, acid-fast, rod-shaped organisms. Three species, *Mycobacterium fortuitum*, *Mycobacterium marinum*, and *Mycobacterium chelonae*, are pathogenic to tropical- and temperate-climate fish [36]. All freshwater and saltwater fish are susceptible, and there is a zoonotic potential [37]. Generally, in fish, the infection is a chronic wasting disease. There is emaciation, a sunken abdomen, ascites, external ulcerations or nodules, deformities of the spinal column, exophthalmia, and changes in coloration with lifting of the scales. Internal organs generally support multiple miliary granulomas [36]. On exfoliative cytology, there is a mixed inflammatory response, with the granulomas supporting dense capsular proliferations of elongate epithelioid cells, which are epithelioid macrophages or the lining capsule of fibrocytes. An acid-fast stain can readily identify these rod-shaped bacteria within the cytoplasm of histiocytic cell types (Fig. 10).

There are several fungal infections that have been described of fish. Saprolegnia is widespread in freshwater and brackish water species. These infections are generally secondary to poor management or chronic parasitic infestation. On exfoliative cytology, the fungal organisms are long-branched, aseptate, hyphal structures (Figs. 11 and 12). Other fungal infections include those that are caused by oomycetes, which include the aphanomycete and deuteromycete groups. These infections can result in deep dermal ulcers that may extend down to the underlying bones. Fungal infections are also typically secondary disease conditions to immunosuppression from recent stressors. On an exfoliative cytology smear, there are numerous septate, branched, pigmented fungal hyphae [38].

Fusarium infections have been reported in several species of fish and other marine life, including a loggerhead sea turtle and a sea snake [38–40]. This group of fungal organisms is suspected to represent environmental contaminants, and suboptimal environmental temperature is a contributing factor to

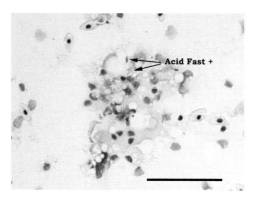

Fig. 10. Exfoliate touch preparation of an ulcerative skin lesion. Within the histiocytes, there are intracytoplasmic acid-fast positive bacteria consistent with *Mycobacterium*. Acid-fast stain. Bar = 0.05 mm.

the development of disease. In fish, it can result in large, necrotic, friable lesions through the skin to the level of the underlying musculature. It is suspected that these lesions start as some type of trauma and are subsequently colonized by the *Fusarium* species. On cytology, there are inflammatory cells with hemorrhage and a large number of fungal organisms. The fungal hyphae are typically characterized by nonparallel cell walls with irregular internal septations and dichotomous branching. Concurrent bacterial infections are common. With culture of these organisms, they develop the classic canoe-shaped or banana-shaped microconidia.

There are reports of algal-induced dermatitis of fish, most recently in ornamental cichlids (*Pseudotropheus zebra*) [41]. Such infections have been identified primarily on the gills and skin and, occasionally, within the visceral tissues. The common algal groups associated with infections of fish are the Chlorococcales. These include *Chlorella* and *Chlorochytrium* species,

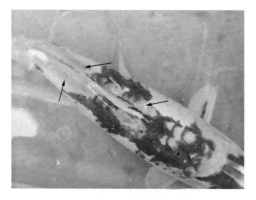

Fig. 11. Cotton-like fungal growths (*arrows*) on a koi carp.

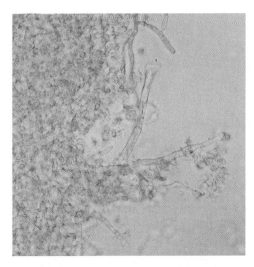

Fig. 12. Saprolegnia mycelial mat. Wet mount. Oil (original magnification ×100) objective.

both of which have been identified on fish. These may result in granuloma formations by a fibrous connective tissue capsule surrounding a large number of irregular oval to circular cells. These are nonfilamentous structures.

Clinical signs of external parasitism in fish include flashing (rapid movement by fish on objects in a tank or pond or onto the floor of a pond or tank), gasping at the surface, increased opercular movements, clamped fins, decreased appetite, lethargy, and depression. A wet mount examination of the slime layer can be used to diagnose these parasites. Most common parasites are identified by their characteristic size and movement. Wet mount preparations should be evaluated promptly, because some parasites may die on the slide or be susceptible to desiccation as the material dries. This is particularly important when performing wet mount preparations pond side.

*Pleistophora* is a microsporidian parasite that has a direct life cycle. This organism commonly causes a disease known as neon tetra disease. It does affect a wide range of tropical freshwater fish [42]. On clinical examination, the fish has small ulcerative lesions along the lateral body wall in the vicinity of the lateral line. A deep scraping and cytologic examination of the affected areas may identify spherical cysts embedded within the muscle tissue. These spherical cysts (xenomas) are typically composed of numerous smaller membrane-bound packets that are filled with oval spores.

*Ichthyophthirius multifiliis* affects freshwater fish worldwide and can be virulent under aquarium conditions [6,7]. The saltwater counterpart is *Cryptokaryon irritans*. They have a complex life cycle that includes stages on the host and in the environment. Clinically, they appear as small white spots on the body of the fish. These spots are called trophonts, which are the encysted feeding stages. The trophont enlarges and breaks through the epithelium to drop to the bottom of the aquarium. The organism is now a tomont, which divides into hundreds of free-living ciliated theronts. The theronts penetrate the skin and gill epithelium to start the cycle again. Finding the encysted ciliate (the trophont) within the epithelium on exfoliate cytology is diagnostic. *I multifiliis* is a holotrich ciliated parasite with a variable size and shape of the trophonts and with a large C-shaped macronucleus (Fig. 13). The protozoa may also serve as a vector for pathogenic bacteria, such as *A hydrophila*, providing a portal of entry when it traumatizes the skin [43].

*Ichthyobodo necator* is a flagellated parasite previously called *Costia necatrix* [6]. There are closely related protozoal species that infect marine fish [44]. The organism feeds directly on epithelial cells of the body and gills. Respiratory distress is a common sign. It is a disease more commonly seen by wholesalers and retailers. Usually, the flagellate protozoon is attached to the gill or skin epithelium and can be identified by its movement, which resembles a flickering flame. It is about the size of fish red blood cells.

Many trichodinid species infest marine and freshwater fish [6,7,45]. They all have a similar morphology, which includes cilia, a circular shape when viewed from the top of the parasite, and a ring of hook-like denticules (Fig. 14). The parasites are commonly found on skin and gills of fish from waters with high organic loads. With a heavily infested fish, their adherence and suction on the epithelium may cause enough damage to produce the clinical signs of anorexia and weight loss.

*Epistylis* and a related species, *Vorticella,* are stalked ciliated protozoans generally found attached to vegetation or crustaceans (Fig. 15). In high

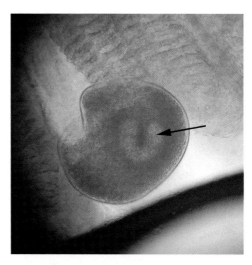

Fig. 13. *Ichthyophthirius multifiliis* on a gill preparation. The arrow is on the C-shaped macronucleus. Wet mount.

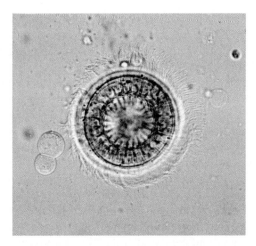

Fig. 14. Trichodina. Wet mount. Oil (original magnification ×100) objective.

organic water, they proliferate and attach to fish and eggs. They frequently affect goldfish and many species of bottom-dwelling freshwater fish. Both have been isolated from ulcerated areas in association with *Aeromonas* bacteria [2].

The trematodes *Dactylogyrus* spp and *Gyrodactylus* spp are common parasites of freshwater and saltwater fish [6,7]. *Dactylogyrus* spp are usually associated with the gills [6]. They can cause gill hyperplasia, destruction of gill epithelium, and clubbing of a gill's filaments, resulting in hypoxia. Clinical

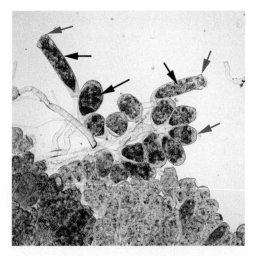

Fig. 15. Epistylis. The stalks are supporting zooids (*black arrows*) that have cilia surrounding the mouth (*red arrows*). Wet mount. Oil (original magnifiction ×100) objective.

signs are evident in closed aquarium systems in which the fish may be exposed to a heavy parasite load. *Gyrodactylus* spp commonly infest the body of the fish (Fig. 16) [6]. This group is viviparous, and embryos can be seen inside the adults. The parasites feed on blood and epithelium, causing localized hemorrhagic areas, excessive mucus, and ulcers. *A hydrophila* has been isolated from *Gyrodactylus* spp, indicating that they may actively transmit the bacteria.

*Lernaea* species are parasitic copepods commonly called "anchor worms," and they can be found on goldfish, koi, and other freshwater fish [6,7]. The embedded head of the female worm forms an anchor in the tissues of the body, fins, or gills. The mature female parasite produces egg sacs that trail out of the attachment site. The parasite causes irritation and localized reactions, with secondary bacterial infections. Even when it is removed, focal areas of dermal hyperplasia or fibrosis may persist.

*Argulus* is the fish louse commonly found in goldfish and koi production ponds [6]. These branchurian crustaceans range in size from 5 to 8 mm. They are readily identified by their suckers, flattened appearance, and shell-like carapace (Figs. 17 and 18). The organism feeds on the blood of the fish by means of a stiletto. The sting may result in a severe hemorrhagic reaction that frequently becomes infected with secondary bacterial infections [46].

Skin tumors are fairly common in fish, with neurofibromas and fibromas being the most common. These masses must be differentiated from lesions of hyperplasia, which can result in raised lesions similar to many of the tumors as well as masses of granulomas or sites of parasitic attachment, such as from *Lernaea* and related species.

Spindle-cell tumors are common skin tumors of fish [3]. This general classification is based on the histologic morphology of the elongate to spindle-shaped neoplastic cells arranged in streaming bundles. The group

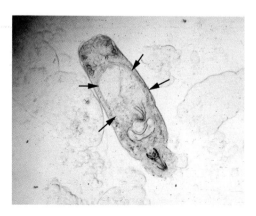

Fig. 16. Gyrodactylus with anchors and an embryo (*arrows*). Wet mount.

Fig. 17. Argulus (*arrow*) on the tail fin.

of mesenchymal tumors includes benign schwannomas, fibromas, and neurofibromas as well as malignant schwannomas, fibrosarcomas, and neurofibrosarcomas. In most cases, further ultrastructural studies are necessary to obtain a more specific classification. These tumors are generally tan to red, raised, firm nodules that are occasionally pedunculated. On cytologic aspirates, if cells are identified, they are spindle shaped with variable amounts of a light blue cytoplasm and indistinct borders. They are arranged individually or in large woven mats. Their cell nuclei are round to oval with clumped chromatin and multiple indistinct nucleoli. Surgical removal is the only therapy described. These tumors are locally infiltrative, and recurrence is common.

There are several pigment cell tumors that are described in a variety of fish. Melanomas and chromatophoromas as well as chromatoblastomas have been described in swordfish, platyfish (*Xiphophorus maculates*), freshwater drum, croakers, koi, and goldfish as well as in damsel fish [3,47,48].

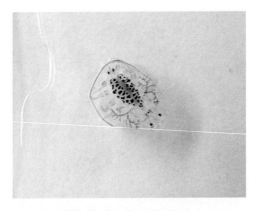

Fig. 18. Argulus wet mount.

These tumors are composed of pigmented cells that are typically pleomorphic (stellate, spindle shaped, or dendritic).

Other tumors that might present with cutaneous lesions include malignant lymphoma and lipomas [49]. Malignant lymphoma has been described in several fish species [3,50]. In northern pike and muskellunge, malignant lymphoma is suspected to be of retroviral origin [3]. The cause in other fish species has not been determined. These can present as systemic disease processes involving multiple organs or focal areas of masses on the skin. The cytology preparation is similar to that of other animal species. Generally, these are densely cellular preparations with large numbers of round cells that have a scant amount of cytoplasm and a round cell nucleus having a clumped chromatin pattern and variably prominent nucleoli. Cytologically, lipomas are preparations of lipid characterized by clear vacuoles on the stained slides.

Xanthomas have recently been described in eels (wolf eel [*Anarrhichthys ocellatus*], California moray eel [*Gymnothorax mordax*], and a snowflake eel [*Echidna nebulosa*]) [51]. Xanthomas are not neoplastic, although they are locally invasive mass lesions. These masses are composed of foamy macrophages and cholesterol clefts (Fig. 19).

*Kidney*

The kidney is divided into two unique portions: the head (anterior) kidney and the tail (posterior) kidney. The head kidney, located dorsally and caudal to the gill arches, is the major lymphoid and hematopoietic organ of the fish. The posterior kidney is located dorsal to the swim bladder and is the site of renal excretion and adrenal hormone secretion (interrenal tissue). The kidney is the organ of choice for culture and isolation of bacteria causing systemic infections.

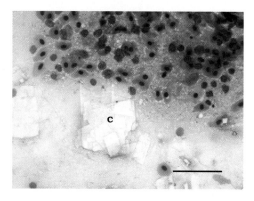

Fig. 19. Exfoliative cytology of a xanthoma from an eel. The cholesterol crystals (C) are irregular, angular, light blue to clear structures. May-Grunwald Giemsa stain. Bar = 0.05 mm.

The kidney is important for retaining salts and excreting water in freshwater fish. These fish have an internal salt concentration that is greater than that of the surrounding environment. Freshwater fish drink little to no water and produce urine that is hypotonic to the blood. Their kidneys have numerous glomeruli. Marine fish have an internal salt concentration lower than that of their environment. They tend to lose water by osmosis and gain salts by diffusion. Marine fish drink large quantities of water to compensate for water loss through the gills. They have renal modifications to minimize water loss, such as few to no glomeruli in some cases.

Several disease conditions can be evaluated by cytology. On tissue squash preparations of the kidneys, bacterial granulomas can be readily identified and are characterized by central cores of granular debris surrounded by concentric layers of epithelioid macrophages and fibrocytes. Fungal and protozoal infections can also be identified (Fig. 20).

Polycystic kidney disease has been described in goldfish. Grossly, the kidneys are distorted and greatly expanded with numerous cysts (Fig. 21). There are several etiologies proposed for the lesions. The mxyozoan *Hoeferellus carassii* (formerly known as *Mitraspora cyprini*) causes massive renal hypertrophy specifically of the renal tubules and ureters [52]. Even late in the disease, some spores and sporoblast remnants may be found within the cystic spaces. Another form of polycystic kidney disease focuses on the glomeruli and results in the cysts forming from extreme dilation of the glomeruli. An etiology has not been determined [53]. Feral goldfish from a polluted harbor also developed polycystic kidneys that did not seem to involve the

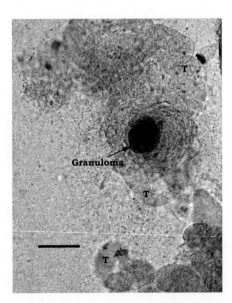

Fig. 20. Kidney from a betta. The renal tubules (T) are prominent. There is a bacterial granuloma present. Wet mount. Bar = 0.1 mm.

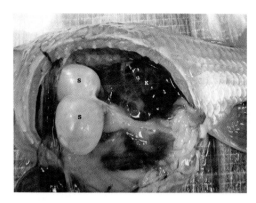

Fig. 21. Gross finding of polycystic renal disease in a goldfish. K, kidney; S, swim bladder.

glomeruli or tubules; however, these fish also had a significant number of renal interstitial bacterial granulomas [54]. In general, these cysts should be aspirated for cytologic analysis and evaluation for parasites. In most cases, the fluid has low numbers (or none) of cells and is low in protein content. Occasionally, the cysts can rupture, and a hemorrhagic fluid is aspirated.

*Liver*

The liver is yellow brown to dark red in coloration and is lobed. It functions as a mammalian liver does. Pancreatic tissue may be identified in some species, and most fish have melanomacrophage centers (MMCs) [55]. These normal lymphomyeloid structures are also found in the spleen, liver, and kidney. Cytologically, MMCs are nodular masses of brown to black cells that are large and round to polyhedral. The cytoplasmic pigment, which ranges from brown to black, is within a foamy cytoplasm. The exact function of the MMCs is not completely determined, but they are suspected to be involved with antigen processing. The pigments within the cells also include lipofuscin, which is a complex of polyunsaturated fatty acids, hemosiderin from hemoglobin degradation, and melanin, which acts as a free radical scavenger [55].

The liver can be involved in systemic infections, although the kidney is more typically the organ of choice for systemic infectious diseases. The liver is the common site for granulomas of *Mycobacterium* species, *Nocardia* species [56,57], encysted nematodes, and metacercariae of digenetic trematodes (Fig. 22).

Fatty hepatic degeneration is a common lesion in captive fish that have been on high-fat diets or overfed [58]. These fish may present as overconditioned animals, usually with a firmly distended coelom. On aspiration cytology or squash preparations, the hepatocytes have multiple to coalescing clear vacuoles in the cytoplasm (Fig. 23).

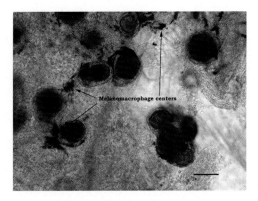

Fig. 22. Liver from a betta. A large number of granulomas are present. Most are associated with the melanomacrophage centers. Wet mount. Bar = 0.1 mm.

Hepatic tumors may be identified on cytology; however, unless the change is diffuse throughout the liver, aspiration cytology may not capture the neoplastic cells. Usually, these fish have concurrent ascitic fluid.

The gallbladder, a sac-like structure embedded in the liver, may be small or large depending on the species and any degree of anorexia. Myxosporean species are frequently found in the gallbladder of saltwater fish [59,60] and, occasionally, in goldfish [61]. Myxosporean species are amoeboid when immature (plasmodia) and become more disk shaped and elongated as they age [60]. Generally no histologic lesions are present; the amoeboid structures (up to 3 cm in size) and spores (approximately 8–25 μm) may be free-floating within the gallbladder. Identifying the polar capsules (more easily identified with Giemsa or Wright staining) of the spores confirms a myxozoan infection [2].

The protozoan *Spironucleus* (Hexamita) can also be found in the bile fluid [19,20]. This is described in the section on the intestinal tract.

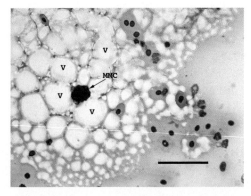

Fig. 23. Impression smear of hepatocytes with large, discrete, cytoplasmic vacuoles (V). A melanomacrophage center (MMC) is present. Modified Giemsa stain. Bar = 0.05 mm.

## Reproductive tract

The reproductive tract can be evaluated by cytologic samples collected as a percutaneous aspirate of masses or by impression preparations made during surgical exploration or necropsy. Percutaneous aspiration can be performed with careful palpation and isolation of the mass or an ultrasound-guided process. Neoplasia and infections attributable to bacteria, fungus, or protozoa organisms may be appreciated by cytologic examination.

Although coccidian, microsporidian, and myxosporean protozoal parasites are described in the gonads of economically important food fish [59,62], there are few documented cases in aquarium fish [63]. If these parasites are present in aquarium fish species, aspiration cytology or a squash preparation of the gonadal tissue should provide a diagnosis. Systemic bacterial and fungal infections, especially if they result in granuloma formations, can be recognized. It is important not to confuse ovarian follicles with granulomas (Fig. 24).

Several gonadal tumors are described in food or sport fish [50,64–66]. Koi have a significant incidence of ovarian tumors [67,68]. These tumors can be massive, and aspiration for cytology is useful in differentiating these tumors from retained eggs or inflammatory lesions. The cytologic appearance of these tumors has not been reported; however, histologically, they are composed of pleomorphic cells.

Seminomas have been reported in several aquaria fish [21,69,70]. These produce a large intracoelomic mass. The cytology preparation is densely cellular, primarily with large numbers of densely packed cells. These may be difficult to differentiate from malignant lymphomas.

## Respiratory system

Nitrogenous waste excretion is primarily the function of the gills rather than the kidneys. Up to 75% of ammonia wastes (main excretory product

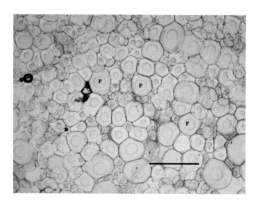

Fig. 24. Ovarian follicles (F). Wet mount. Bar = 0.5 mm.

in fish) are removed by the gills. Chloride cells in the gill are responsible for osmoregulation and mineral balance. The gills can be damaged by extreme pH fluctuations; high ammonia levels in the water; such parasites as *Ichthyophthirius, Trichodina, Chilodonella, Ichthyobodo (Costia)* (Fig. 25), *Scyphydia, and Oodinium*; mongenetic trematodes; metacercaria; bacteria; viruses; and toxins (eg, chlorine of untreated tap water, copper). The gills' response to damage is excessive production of mucus, which inhibits gas exchange. Necrosis of the gills occurs in focal patches or diffusely with KHV (CyHV-3) infection [24].

The general gill structure provides a large surface area for gas exchange and a countercurrent exchange that results in the availability of approximately 80% of the dissolved oxygen. The gill surface consists of a thin epithelium carried on lamellae. Teleost fish have four pairs of gill arches supported by a bony skeleton. Two hemibranches (filaments or primary lamellae) arise from the arch on the caudal edge. The free tips of the hemibranches on each arch touch those of the adjacent arch. The hemibranch filaments are long and thin. On the dorsal and ventral surfaces of the hemibranch filaments are semicircular folds of epithelium, the secondary lamellae. These are the sites of gas exchange.

On gill clip preparations, normal gills appear reddish pink and the primary lamellae are clearly visible. If the gills appear swollen and edematous with excessive mucus on the surface and there is fusing of the primary lamellae, this is a nonspecific change that can be attributable to any number of gill irritants. Causes of branchitis include parasites, bacteria, increased suspended solids, and toxins. Generally, the secondary lamellae are shortened and rounded and are occasionally fused.

Parasites can readily be identified on the gill wet mounts. There may be congestion, inflammatory cells, and excess mucus production associated with the parasites. Review the parasites listed in the section on skin cytology, because many can also be recognized on the gills.

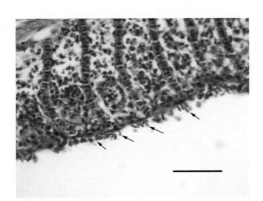

Fig. 25. Numerous *Ichthyobodo necator* (*arrows*), a flagellated parasite, are attached to the gill epithelium. Hematoxylin-eosin section. Bar = 0.05 mm.

## Swim bladder

Buoyancy problems typically characterize lesions within the swim bladder. The fish may float to the surface, sink to the bottom, or maintain an abnormal position in the water column. Some of the common diseases that might be identified by cytology include fungal and bacterial infections, parasites, and neoplasms. Few diseases are recognized in aquarium fish, with the exception of overinflation of the bladder, which is usually an idiopathic condition.

Radiographically, it should be possible to identify if the swim bladder is distended with a gas or fluid. Aspiration through the body wall into the swim bladder can be performed for fluid analysis. Radiographs or ultrasonography can be helpful in directing the aspiration needle. Aerocystitis is characterized by an inflammatory response, and inflammatory cells and possible causative agents may be recognized (Fig. 26).

In cultured and commercial fish, several swim bladder diseases are recognized. These may be recognized in fish on public display or wild caught for the pet trade. Infective pentastome larvae, *Sebekia wedli*, can be encapsulated or free-living in the swim bladders of cichlid fishes, *Tilapia rendalli*, and *Oreochromis mossambicus* [71]. These infections are believed to result in smaller and lighter fish compared with nonparasitized fish. Nematodes in the intestine and swim bladder of the European eel, *Anguilla anguilla*, is a prevalent problem [72,73]. Several myxosporean parasites involve the swim bladder, including *Sphaerospora renicola*, a parasite causing renal sphaerosporosis in common carp (*C carpio*) and *Kudoa iwatai* in wild and cultured fish in the Red Sea [74]. *S renicola* accumulate in the swim bladder wall, causing hemorrhages and necrosis [75]. The swim bladder is also secondarily involved with systemic infections, such as from *A hydrophila* [76] and epizootic hematopoietic necrosis virus (EHNV) infection [77]. Tumors of the swim bladder are rare; however, an outbreak of leiomyosarcoma in

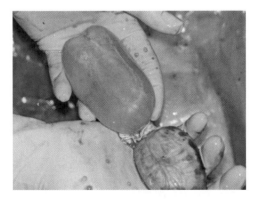

Fig. 26. Fluid-filled swim bladder from a koi. Mycobacteria were isolated from the fluid. (Courtesy of B. Szignarowitz, DVM, Rescue, CA)

the swim bladders of Atlantic salmon has been described, and a novel piscine retrovirus has been identified in association with the tumors [78].

## Thyroid gland

Thyroid hyperplasia as well as thyroid adenoma or carcinoma presents as large bilateral nodular swellings at the base of the gills and extending along the lower gill arches. These masses distend the opercula. These have been described in fish and sharks [79,80]. It is suspected that in marine fish and sharks, there may be exposure to dietary goitrogenic substances, although many factors may influence thyroid growth [81]. The thyroid gland is generally a diffuse organ located along the floor of the gill chamber as well as in the spleen, heart, and cranial kidney. On cytology, there are variable amounts of a homogeneous colloid staining grayish purple, with possible clusters of epithelial cells. Many times, the results of aspiration cytology are poorly cellular within only occasional uniform-appearing epithelial cells. Aspiration would be helpful in differentiating between inflammation tissue and hyperplasia or neoplasia.

## Summary

Cytology is an essential part of a diagnostic workup in cases of aquatic animal diseases. It is simple to perform, inexpensive, and can yield quick and valuable results. External parasites, bacterial and fungal diseases, and gastrointestinal infestations are easily determined with wet mount cytology. Because of the relatively small number of nonlethal diagnostic techniques available for aquatic species (although this number is slowly rising), cytologic testing should be considered in every case. Early diagnosis can lead to more effective treatment plans, ensuring a better prognostic outcome in our patients.

## Acknowledgments

The authors thank Bronwyn Szignarowitz, DVM, for her review of the work and valuable suggestions. They also thank the practitioners from the Veterinary Information Network fish classes, who have helped to mold the techniques and value of the procedures.

## References

[1] Chitwood M, Lichtenfels JR. Identification of parasitic metazoa in tissue sections. Exp Parasitol 1972;32(3):407–519.
[2] Noga EJ. Fish disease: diagnosis and treatment. St. Louis (MO): Mosby; 1996.
[3] Groff JM. Neoplasia in fishes. Vet Clin North Am Exot Anim Pract 2004;7(3):705–56.

[4] Reavill DR. Common diagnostic and clinical techniques for fish. Vet Clin North Am Exot Anim Pract 2006;9(2):223–35.
[5] Marselas GA, Stoskopf MK, Brown MJ, et al. Abdominal ascites in electric eels (Electrophorus electricus) associated with hepatic hemosiderosis and elevated water pH. J Zoo Wildl Med 1998;29(4):413–8.
[6] Thilakaratne ID, Rajapaksha G, Hewakopara A, et al. Parasitic infections in freshwater ornamental fish in Sri Lanka. Dis Aquat Organ 2003;54(2):157–62.
[7] Kim JH, Hayward CJ, Joh SJ, et al. Parasitic infections in live freshwater tropical fishes imported to Korea. Dis Aquat Organ 2002;52(2):169–73.
[8] Rigby MC, Font WF, Deardorff TL. Redescription of Camallanus cotti Fujita, 1927 (Nematoda: Camallanidae) from Hawaii. J Parasitol 1997;83(6):1161–4.
[9] Muench TR, White MR. Cryptosporidiosis in a tropical freshwater catfish (Plecostomus spp.). J Vet Diagn Invest 1997;9(1):87–90.
[10] O'Donoghue PJ. Cryptosporidium infections in man, animals, birds and fish. Aust Vet J 1985;62(8):253–8.
[11] Reavill DR, Murray MJ, Hoech J. Intestinal cryptosporidiosis in Knysna Seahorse (Hippocampus capensis). In: Proceedings of the International Association for Aquatic Animal Health. Galveston (TX): 2004. p. 23.
[12] Alvarez-Pellitero P, Sitja-Bobadilla A. Cryptosporidium molnari n. sp. (Apicomplexa: Cryptosporidiidae) infecting two marine fish species, Sparus aurata L. and Dicentrarchus labrax L. Int J Parasitol 2002;32(8):1007–21.
[13] Hoffman GL. Eimeria aurati n. sp. (Protozoa: Eimeriidae) from goldfish (Carassius auratus) in North America. J Protozool 1965;12(2):273–5.
[14] Landsberg JH, Paperna I. Intestinal infections by Eimeria (s. l.) vanasi n. sp. (Eimeriidae, Apicomplexa, Protozoa) in cichlid fish. Ann Parasitol Hum Comp 1987;62(4):283–93.
[15] Upton SJ, Stamper MA, Osborn AL, et al. A new species of Eimeria (Apicomplexa, Eimeriidae) from the weedy sea dragon Phyllopteryx taeniolatus (Osteichthyes: Syngnathidae). Dis Aquat Organ 2000;43(1):55–9.
[16] Solangi MA, Overstreet RM. Biology and pathogenesis of the coccidium Eimeria funduli infecting killifishes. J Parasitol 1980;66(3):513–26.
[17] Sangmaneedet S, Smith SA. In vitro studies on optimal requirements for the growth of Spironucleus vortens, an intestinal parasite of the freshwater angelfish. Dis Aquat Organ 2000; 39(2):135–41.
[18] Paull GC, Matthews RA. Spironucleus vortens, a possible cause of hole-in-the-head disease in cichlids. Dis Aquat Organ 2001;45(3):197–202.
[19] Jorgensen A, Sterud E. SSU rRNA gene sequence reveals two genotypes of Spironucleus barkhanus (Diplomonadida) from farmed and wild Arctic charr Salvelinus alpinus. Dis Aquat Organ 2004;62(1–2):93–6.
[20] Sterud E, Mo TA, Poppe TT. Systemic spironucleosis in sea-farmed Atlantic salmon Salmo salar, caused by Spironucleus barkhanus transmitted from feral Arctic char Salvelinus alpinus? Dis Aquat Organ 1998;33(1):63–6.
[21] Hubbard GB, Fletcher KC. A seminoma and a leiomyosarcoma in an albino African lungfish (Protopterus dolloi). J Wildl Dis 1985;21(1):72–4.
[22] Pliss GB, Khudoley VV. Tumor induction by carcinogenic agents in aquarium fish. J Natl Cancer Inst 1975;55(1):129–36.
[23] Waltzek TB, Kelley GO, Stone DM, et al. Koi herpesvirus represents a third cyprinid herpesvirus (CyHV-3) in the family. J Gen Virol 2005;86(Pt 6):1659–67.
[24] Szignarowitz B. Update on Koi herpesvirus. Exotic DVM Magazine 2005;7(30):92–5.
[25] Majeed SK, Douglas M. Pox-like lesions in a colony of carp (Cyprinus carpio). Vet Q 1986; 8(1):76–80.
[26] Lewbart GA. Clinical snapshot. Compend Contin Educ Pract Vet 2000;22(10):922.
[27] Paperna I, Vilenkin M, de Matos AP. Iridovirus infections in farm-reared tropical ornamental fish. Dis Aquat Organ 2001;48(1):17–25.

[28] Sudthongkong C, Miyata M, Miyazaki T. Iridovirus disease in two ornamental tropical freshwater fishes: African lampeye and dwarf gourami. Dis Aquat Organ 2002;48(3): 163–73.
[29] Gibson-Kueh S, Netto P, Ngoh-Lim GH, et al. The pathology of systemic iridoviral disease in fish. J Comp Pathol 2003;129(2–3):111–9.
[30] Reavill DR. Bacterial diseases of ornamental fish. Seminars in Avian and Exotic Pet Medicine 1993;2(4):179–83.
[31] Kozinska A, Figueras MJ, Chacon MR, et al. Phenotypic characteristics and pathogenicity of Aeromonas genomospecies isolated from common carp (Cyprinus carpio L.). J Appl Microbiol 2002;93(6):1034–41.
[32] Wiklund T, Dalsgaard I. Occurrence and significance of atypical Aeromonas salmonicida in non-salmonid and salmonid fish species: a review. Dis Aquat Organ 1998;32(1):49–69.
[33] Corral FD, Shotts EB, Brown J. Adherence, haemagglutination and cell surface characteristics of motile aeromonads virulent for fish. J Fish Dis 1990;13:255–68.
[34] Ormen O, Granum PE, Lassen J, et al. Lack of agreement between biochemical and genetic identification of Aeromonas spp. APMIS 2005;113(3):203–7.
[35] Dror M, Sinyakov MS, Okun E, et al. Experimental handling stress as infection-facilitating factor for the goldfish ulcerative disease. Vet Immunol Immunopathol 2006;109(3–4): 279–87.
[36] Decostere A, Hermans K, Haesebrouck F. Piscine mycobacteriosis: a literature review covering the agent and the disease it causes in fish and humans. Vet Microbiol 2004;99(3–4): 159–66.
[37] Lehane L, Rawlin GT. Topically acquired bacterial zoonoses from fish: a review. Med J Aust 2000;173(5):256–9.
[38] Yanong RP. Fungal diseases of fish. Vet Clin North Am Exot Anim Pract 2003;6(2): 377–400.
[39] Reavill DR, Crow GL, Okimoto B. Fungal dermatitis in a yellow-bellied sea snake (Pelamis platurus). In: Proceedings of the International Association for Aquatic Animal Medicine. Galveston (TX); 2004. p. 38–9.
[40] Cabanes FJ, Alonso JM, Castella G, et al. Cutaneous hyalohyphomycosis caused by Fusarium solani in a loggerhead sea turtle (Caretta caretta L.). J Clin Microbiol 1997;35(12): 3343–5.
[41] Yanong RP, Francis-Floyd R, Curtis E, et al. Algal dermatitis in cichlids. J Am Vet Med Assoc 2002;220(9):1353–8.
[42] Lom J, Nilsen F. Fish microsporidia: fine structural diversity and phylogeny. Int J Parasitol 2003;33(2):107–27.
[43] Liu YJ, Lu CP. Role of Ichthyophthirius multifilis in the infection of Aeromonas hydrophila. J Vet Med B Infect Dis Vet Public Health 2004;51(5):222–4.
[44] Todal JA, Karlsbakk E, Isaksen TE, et al. Ichthyobodo necator (Kinetoplastida)—a complex of sibling species. Dis Aquat Organ 2004;58(1):9–16.
[45] Colorni A, Diamant A. Hyperparasitism of trichodinid ciliates on monogenean gill flukes of two marine fish. Dis Aquat Organ 2005;65(2):177–80.
[46] Bandilla M, Valtonen ET, Suomalainen LR, et al. A link between ectoparasite infection and susceptibility to bacterial disease in rainbow trout. Int J Parasitol 2006;36(9): 987–91.
[47] Ishikawa T, Masahito P, Matsumoto J, et al. Morphologic and biochemical characterization of erythrophoromas in goldfish (Carassius auratus). J Natl Cancer Inst 1978;61(6):1461–70.
[48] Etoh H, Hyodo-Taguchi Y, Aoki K, et al. Incidence of chromatoblastomas in aging goldfish (Carassius auratus). J Natl Cancer Inst 1983;70(3):523–8.
[49] McCoy CP, Bowser PR. Lipoma in Channel catfish (Ictalurus punctatus Rafinesque). J Wildl Dis 1985;21(1):74–8.
[50] Earnest-Koons KA, Schachte JH Jr, Bowser PR. Lymphosarcoma in a brook trout. J Wildl Dis 1997;33(3):666–9.

[51] Reavill DR, Adams L, Hoech J, et al. Xanthomatous lesions in three eels. In: Proceedings of the International Association for Aquatic Animal Medicine. Nassau (Bahamas): 2006. p. 85.
[52] Ahmed ATA. Kidney enlargement disease of goldfish in Japan. Jpn J Zool 1974;17:37–65.
[53] Schlumberger CJ. Polycystic kidney (mesonephros) in the goldfish. Arch Pathol 1950;50: 400–10.
[54] Munkittrick KR, Moccia RD, Leatherland JF. Polycystic kidney disease in goldfish (Carassius auratus) from Hamilton Harbour, Lake Ontario, Canada. Vet Pathol 1985;22(3):232–7.
[55] Agius C, Roberts RJ. Melano-macrophage centres and their role in fish pathology. J Fish Dis 2003;26(9):499–509.
[56] Conroy DA. Nocardiosis as a disease of tropical fish. Vet Rec 1964;76:676.
[57] Valdez IE, Conroy DA. The study of a tuberculosis-like condition in neon tetras (Hyphessobrycon Innesi). II. Characteristics of the bacterium isolated. Microbiol Esp 1963;16: 249–53.
[58] Speare DJ. Liver diseases of tropical fish. Semin Avian Exotic Pet Med 2000;9(3):174–8.
[59] Timi JT, Sardella NH. Myxosporeans and coccidians parasitic on engraulid fishes from the coasts of Argentina and Uruguay. Parasite 1998;5(4):331–9.
[60] Diamant A, Whipps CM, Kent ML. A new species of Sphaeromyxa (Myxosporea: Sphaeromyxina: Sphaeromyxidae) in devil firefish, Pterois miles (Scorpaenidae), from the northern Red Sea: morphology, ultrastructure, and phylogeny. J Parasitol 2004;90(6):1434–42.
[61] Hallett SL, Atkinson SD, Holt RA, et al. A new myxozoan from feral goldfish (Carassius auratus). J Parasitol 2006;92(2):357–63.
[62] Nagel ML, Hoffman GL. A new host for Pleistophora ovariae (Microsporida). J Parasitol 1977;63(1):160–2.
[63] Swearer SE, Robertson DR. Life history, pathology, and description of Kudoa ovivora n. sp. (Myxozoa, Myxosporea): an ovarian parasite of Caribbean labroid fishes. J Parasitol 1999; 85(2):337–53.
[64] Budd J, Schroder JD. Testicular tumors of yellow perch, Perca flavescens (Mitchill). J Wildl Dis 1969;5(3):315–8.
[65] Herman RL, Landolt M. A testicular leiomyoma in a largemouth bass, Micropterus salmoides. J Wildl Dis 1975;11(1):128–9.
[66] Borucinska JD, Harshbarger JC, Bogicevic T. Hepatic cholangiocarcinoma and testicular mesothelioma in a wild-caught blue shark, Prionace glauca (L.). J Fish Dis 2003;26(1):43–9.
[67] Lewbart GA. Reproductive medicine in koi (Cyprinus carpio). Vet Clin North Am Exot Anim Pract 2002;5(3):637–48.
[68] Ishikawa T, Kuwabara N, Takayama S. Spontaneous ovarian tumors in domestic carp (Cyprinus carpio): light and electron microscopy. J Natl Cancer Inst 1976;57(3):579–84.
[69] Weisse C, Weber ES, Matzkin Z, et al. Surgical removal of a seminoma from a black sea bass. J Am Vet Med Assoc 2002;221(2):280–3.
[70] Masahito P, Ishikawa T, Takayama S, et al. Gonadal neoplasms in largemouth bass, Micropterus salmoides and Japanese dace (ugui), Tribolodon hakonensis. Gann 1984;75(9): 776–83.
[71] Junker K, Boomker J, Booyse DG. Pentastomid infections in cichlid fishes in the Kruger National Park and the description of the infective larva of Subtriquetra rileyi n. sp. Onderstepoort J Vet Res 1998;65(3):159–67.
[72] Pilecka-Rapacz M, Sobecka E. Nematodes of the intestine and swim bladder of the European eel Anguilla anguilla (L.) ascending Pomeranian rivers. Wiad Parazytol 2004;50(1): 19–28.
[73] Norton J, Rollinson D, Lewis JW. Epidemiology of Anguillicola crassus in the European eel (Anguilla anguilla) from two rivers in southern England. Parasitology 2005;130(Pt 6): 679–86.
[74] Diamant A, Ucko M, Paperna I, et al. Kudoa iwatai (Myxosporea: Multivalvulida) in wild and cultured fish in the Red Sea: redescription and molecular phylogeny. J Parasitol 2005; 91(5):1175–89.

[75] Molnar K. The occurrence of Sphaerospora renicola K-stages in the choroidal rete mirabile of the common carp. Folia Parasitol (Praha) 1993;40(3):175–80.
[76] Topic-Popovic N, Teskeredzic E, Strunjak-Perovic I, et al. Aeromonas hydrophila isolated from wild freshwater fish in Croatia. Vet Res Commun 2000;24(6):371–7.
[77] Reddacliff LA, Whittington RJ. Pathology of epizootic haematopoietic necrosis virus (EHNV) infection in rainbow trout (Oncorhynchus mykiss Walbaum) and redfin perch (Perca fluviatilis L). J Comp Pathol 1996;115(2):103–15.
[78] Paul TA, Quackenbush SL, Sutton C, et al. Identification and characterization of an exogenous retrovirus from Atlantic salmon swim bladder sarcomas. J Virol 2006;80(6):2941–8.
[79] Gridelli S, Diana A, Parmeggiani A, et al. Goitre in large and small spotted dogfish, Scyliorhinus stellaris (L.) and Scyliorhinus canicula (L.). J Fish Dis 2003;26(11–12):687–90.
[80] Hoover KL. Hyperplastic thyroid lesions in fish. Natl Cancer Inst Monogr 1984;65:275–89.
[81] Sonstegard R, Leatherland JF. The epizootiology and pathogenesis of thyroid hyperplasia in coho salmon (Oncorhynchus kisutch) in Lake Ontario. Cancer Res 1976;36(12):4467–75.

# Cytologic Diagnosis of Diseases of Invertebrates

Arnaud Van Wettere, DVM, MS[a],*,
Gregory A. Lewbart, MS, VMD, DACZM[b]

[a]*Department of Population Health and Pathobiology, College of Veterinary Medicine, North Carolina State University, 4700 Hillsborough Street, Raleigh, NC 27606, USA*
[b]*Department of Clinical Sciences, College of Veterinary Medicine, North Carolina State University, 4700 Hillsborough Street, Raleigh, NC 27606, USA*

Invertebrates are a vast collection of animals, comprising more than 95% of the earth's species, unified only by the lack of a vertebral column. Ruppert and Barnes [1] state that the invertebrates are a group of unrelated taxa that share no universal "positive" traits. There are currently more than 30 recognized phyla (and there is not universal agreement on what a phylum is in many cases) of invertebrates (not including the protozoans). Many of these phyla are considered "obscure" for no better reason than that they may contain few species or microscopic representatives or have no obvious economic value. In reality, the animals belonging to each phylum are important to the diversity and survival of the planet, even if the group is only studied by a small number of scientists or occupies a restricted habitat. Unfortunately, little is known about the veterinary aspects of many of these taxa, and writing a comprehensive article for all invertebrate phyla would be inefficient currently. Consequently, the authors have elected to describe the general techniques used for collection and interpretation of samples and include examples of diseases that may be diagnosed by cytology in the most commonly encountered or economically important and "visible" metazoan groups. These taxa comprise species kept as pets, in zoos for education, for research, as a food source, or for production of material like silk and pearls. Despite the long-time interest in invertebrate pathology, most studies have focused on pest control or basic physiologic and host response studies. Only limited information is available on diagnostic pathology, with most research being done on economically important species and focusing on population health.

---

* Corresponding author.
  *E-mail address:* arnaud_vanwettere@ncsu.edu (A. Van Wettere).

This article is intended for the exotic animal veterinary practitioner or zoo veterinarian and provides an outline of the use of cytology as an aid to diagnosis of invertebrate diseases in a clinical setting. It should be emphasized that cytologic diagnosis in invertebrate clinical medicine is still in its infancy and more research is necessary to improve the diagnostic methods in the different, and frequently unrelated, invertebrate species. Nevertheless, cytology presents a quick and easy method to evaluate for the presence of infectious and parasitic agents, inflammatory reaction, or neoplasms without killing the animal. The authors hope that this article stimulates clinical veterinarians to collect samples, report results, and establish reference values for normal and disease states of the species they are working with.

## Sample collection and smear preparation

Cytologic samples can be collected as in other species by swabbing, scrapping, imprints, or fine-needle biopsy (Fig. 1). The collection techniques are similar to those described for vertebrates and are not discussed here, with the exception of hemolymph collection.

### Hemolymph collection and smear preparation

Hemolymph of invertebrates is comparable to the blood of vertebrates. Many invertebrates have an open circulatory system in which hemolymph flows to the organs through arteries, passes into the hemocoelomic cavity, and drains to the heart and respiratory organs through progressively larger venous channels [2,3]. In most cases, hemolymph is composed of the copper-containing respiratory pigment hemocyanin, soluble defense molecules (ie, lectins, C-reactive proteins, $\alpha_2$-macroglobulins), and circulating hemocytes [2,4–6]. Hemocytes can be considered as the equivalent of the vertebrate

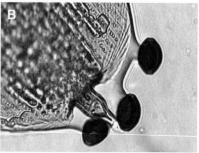

Fig. 1. (*A*) Application of clear transparent tape to the abdomen of a monarch butterfly (*Danaus plexippus*) to evaluate for the presence of *Ophryocystis elektroscirrha* spores. (Courtesy of D. Dombrowski.) (*B*) Squash preparation from transparent tape imprint of monarch butterfly abdomen showing spores of *O. elektroscirrha* on the ectodermal surface of a scale. (Courtesy of A.A. Aguirre and H. Zepeda.)

white blood cells with additional function in coagulation. Hemocytes are involved in phagocytosis and encapsulation of foreign material, coagulation, and, at least in crustaceans, hardening of the cuticle [2]. Hemolymph is the most readily accessible and often relatively easily collected tissue antemortem and can even be obtained from small invertebrates, such as flies and mosquitoes [7].

As a general rule, it is recommended to clean and disinfect the shell or integument at the collection site. In species with a hard cuticle, if the bleeding does not stop rapidly after sample collection, closure of the puncture site can be made using cyanoacrylate or tissue glue, adhesive, or a similar adherent material [8]. Anesthesia may be required for sample collection in some species, such as cephalopods, and the reader is referred to the chapter by Gunkel and Lewbart [9] for additional information. There is wide variation in the anatomy between different species, and the description of hemolymph collection sites is restricted to a few of the more commonly encountered taxa. Hemolymph can be collected from bivalve mollusks from the pericardial region or the sinus of the posterior adductor muscle (Fig. 2) [10,11]. A small notch or hole in the shell can be made with a triangular metal file or drill to facilitate or gain access to the adductor muscle [12]. Hemolymph can be collected from American lobsters by using the ventral abdominal sinus at the junction between the first and second abdominal segments [13]. Hemolymph can be collected from pulmonate snails by entering the hemocoel 5 mm (snails weighing less than 50 g) to 20 mm (snails weighing 200 g) below the pneumostome [14]. Hemolymph can be collected from horseshoe crabs by inserting the needle through the arthrodial membrane between the prosoma and opisthoma into the cardiac sinus (Fig. 3) [15]. Hemolymph can be collected from the heart of spiders; the pericardial sinus is found on the dorsal opisthoma. For additional information on hemolymph collection, the reader is referred to the book edited by Lewbart [16].

Fig. 2. Hemolymph collection from a quahog clam (*Mercenaria mercenaria*). The needle is inserted into the pericardial sinus through the adductor muscle after drilling a small hole through the shell. The hole is closed after collection with epoxy putty.

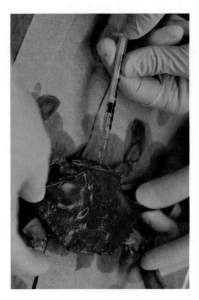

Fig. 3. Hemolymph collection from a blue crab (*Callinectes sapidus*). The needle is inserted in the arthrodial membrane between the prosoma and opisthoma into the cardiac sinus.

Precise information regarding the amount of hemolymph that can be safely collected for diagnostic purposes is rarely available, but it is a well-accepted fact that the loss of "too much" hemolymph can result in death of the animal. Cooper [14,17] writes that loss of a small amount of hemolymph (1% of body weight or less) is generally not harmful and that collection of 10% of the total hemolymph volume (2 mL) in 150- to 200-g African land snails (*Achtina* species) did not have an adverse effect. In horseshoe crabs (*Limulus polyphemus*), in which collection of hemolymph is well documented, collection of approximately 5% of the total hemolymph volume (total hemolymph volume: 25% of wet body weight) is considered safe in healthy individuals [15,18]. The average total blood volume (percentage of wet body weight) of a few arthropod species is given as a guideline: the blue crab (*Callinectes sapidus*) with 25.5%, the European green crab (*Carcinus maenas*) with 33.0%, the rock lobster (*Panulirus longipes*) with 17.8%, the freshwater crayfish (*Cambarus virilis*) with 25.6%, and the forest scorpion (*Heterometrus fulvipes*) with 33.4% [18].

Preparation of hemolymph smears requires slightly different handling than that of vertebrate blood. Hemolymph typically coagulates rapidly after collection and, unfortunately, the most commonly used anticoagulants (eg, ethylenediaminetetraacetic acid [EDTA], heparin, citrate) in clinical practice fail to inhibit coagulation entirely. No matter the methodology used, it is better to follow the general principles discussed in this section. The most important factor is that the collection materials (syringe, needle, and slide) are free of endotoxin, because hemocytes activated by endotoxin respond by degranulating, which results in clotting of the hemolymph and inadequate cell morphology [2,6]. The material used for cell culture is ideal. Hemolymph dilution reduces clotting; therefore, prefilling of the syringe with an anticoagulant or dilution solution before collection of the hemolymph sample may be useful to reduce clotting. Because aquatic invertebrates are osmotic conformers, it is ideal to use an anticoagulant or dilution solution that is iso-osmotic to the ambient environment (ie, sterile sea water) [10]. Finally, working in a cool environment and placing the tube holding the collected hemolymph on ice reduces spreading of hemocytes, and thus decreases clotting and adherence to the collecting tube wall [10].

Several solution and buffer recipes are present in the literature, but they are not practical for clinical veterinarians because they are usually not commercially available and can only be stored for a few days [5,10,11,19]. One practical solution for everyday clinical practice is the suspension of the hemolymph collected in 10% formalin (if possible, a known amount of hemolymph in a known amount of formalin to allow for total hemocyte count [THC]), which stops coagulation and fixes the hemocytes [20]. Formalin fixation, unfortunately, somewhat alters cell morphology and makes cell identification and observation of cellular detail more difficult. Formalin does preserve the sample for long periods, however, and allows processing of the collected samples at a later time [2]. To improve cell concentration, cytocentrifugation (low acceleration at 500 rpm for 10 minutes) of the diluted hemolymph sample is advised [10,13]. The centrifugation or cytospin step is necessary in numerous species because their hemocyte counts are low and a direct smear evaluation would not allow evaluation of a significant number of hemocytes. If preparation of a direct smear is desired, immediate fixation is advised; at least in most crustaceans and bivalve mollusks, direct smears of undiluted hemolymph that dry passively do not allow for fine evaluation of the hemocytes, because the smear tends to dry slowly, resulting in rounded hemocytes that have poor cytoplasmic detail and increased degranulation [10,13]. Messick [21,22] recommends spreading a small drop of hemolymph onto an acid-cleaned poly-L-lysine–coated microscope slide, followed by immediate immersion in fixative solution (Bouin's or Davidson's solution; formalin) for at least 5 minutes and rinsing in running tap water (20 minutes) or 70% ethyl alcohol (until no yellow color remains if Bouin's solution is used) before staining. Not all species necessitate immediate fixation of the smears, and Santarem and Figueras [12] reported successful use of the following protocol in mussels (*Mytilus galloprovincialis*). The hemolymph is gently withdrawn, avoiding any contamination. One hundred microliters of hemolymph is placed onto a clean dry microscope slide and allowed to settle at room temperature (20°C) for 5 to 7 minutes. The excess fluid is carefully removed using a Pasteur pipette, and the slide is immersed vertically into a cytologic fixative solution (2.5% glutaraldehyde in ambient seawater, pH 7.2–7.4). The slide is rinsed in distilled water for 5 minutes, and stained. The stained slide is rinsed in running water and air dried.

## Staining procedures

Smears can be stained with the same stains used for vertebrate medicine, with the most routine stains being Wright-Giemsa stain and Diff-Quick (American Scientific Products, McGaw Park, IL). Diff-Quick is probably the stain most often used by the veterinary clinician and allows somewhat adequate evaluation of the samples. Wright-Giemsa stain is preferred when available, however, for the same reasons as in vertebrate medicine [10,11,23]. Other stains can also be used, with the most common being lactophenol cotton blue stain, Gomori methenamine silver (GMS) stain, periodic acid–Schiff (PAS) stain, Ziehl-Nielsen stain, and Gram stain.

## Sample evaluation and interpretation

Scraping and smears can be evaluated fresh or fixed and stained. Fresh smears are best made using a drop of a watery solution most similar to the environment in which the invertebrate species lives so as to maximize the chance of observing protozoa. After staining and drying, the smear is first evaluated at low magnification (original magnification ×40–×100), as done in vertebrate medicine, for staining quality, cellularity, and large objects like parasites and fungal hyphae. Areas of increased cellularity are evaluated at higher magnification (original magnification ×100–×200), and an idea of the cellular composition is made (inflammatory cells [hemocytes], epithelial cells, spindle cells, and neoplastic cells). To conclude, the smear is evaluated at original magnification ×400 and ×1000 to evaluate and compare individual cells and to search for the presence of microorganisms or parasites [23].

### Organ cytology

Published literature on specific organ cytology is nonexistent or extremely limited. Despite the lack of information, it is still possible to use cytology to help differentiate between inflammation and neoplasia and to demonstrate bacteria, fungal elements, parasites, and, sometimes, viral inclusions (Table 1). As is often the case with unusual species, collection and archiving of samples from healthy specimens are valuable to help with the interpretation of the samples collected from diseased individuals.

### Inflammatory lesion

With the routine Romanowsky-type stain (Wright stain or Diff-Quik), all bacteria stain blue to purple (basophilic), with a few exceptions, such as *Mycobacterium* species, which do not stain but may be observed as negative images [23]. It is often difficult to interpret the results of bacterial or fungal cultures when dealing with unusual species, and cytology results may help to determine the significance of the infectious agent(s) identified [24]. As a general rule, the presence of bacteria associated with inflammatory cells (hemocytes), especially if phagocytized bacteria are observed, indicates a bacterial infection. Extracellular bacteria with no associated inflammation represent bacterial flora or contamination. Determination of the presence of a monomorphic or pleomorphic population of bacteria may also be an important observation. Some invertebrates have significant bacterial flora in their hemolymph, but if bacteria are numerous enough to be visible in a hemolymph smear, this is considered abnormal [2]. The general comment made for bacteria is also valid for fungi. Fungal organisms do not stain well with Romanowsky-type stains but may be seen occasionally. Most often, they do not stain and can be observed as negative images or stained with lactophenol cotton blue stain, PAS, or GMS. As an example, cytology of abdominal contents was part of the examination protocol investigating increased mortality in British wartbiter cricket nymphs and revealed numerous branched fungal hyphae, which, after culture, were revealed to be the insect pathogen *Verticillium lecani* [25]. Parasites may also be observed, because numerous invertebrate species carry parasites, and it must be determined if the parasite observed is pathogenic or commensal. Cytologic evaluation of organ imprints or hemolymph is a well-recognized diagnostic method for several protozoa affecting bivalve mollusks, some of which are reportable diseases (eg, *Marteilia refringens*) (Figs. 4 and 5). Unfortunately, many infectious diseases are detectable in the hemolymph late in the course of the disease, and cytologic diagnosis may be of little help for the individual. As an example, paramoebosis and hematodiniosis are only observable in the hemolymph in the late stages of the disease, because they mainly infect soft tissues [2]. The veterinary clinician also has to consider more unusual possibilities when interpreting cytologic samples, such as symbiont parasites like the mites living near the mouth parts of the hissing cockroach (*Gromphadoryna potentosa*) [26,27]. Parasites using invertebrates as intermediate hosts can also be seen, and it is often not possible to identify them precisely, because larval stages have few to no distinctive features [27]. Human or vertebrate animal pathogens may also be encountered, such as *Cryptosporidium* species oocysts and *Giardia* species cysts, which are possibly carried by shellfish. These represent a potential health issue, even though cryptosporidiosis after the consumption of shellfish has not yet been documented [28]. Examples of diseases possibly diagnosed cytologically are presented in Table 1. For a comprehensive review of the infectious diseases and parasites of shellfish, the reader is referred an article by Bower and colleagues [29]. This text is updated and available on-line [30].

*Neoplasia*

With the exception of a few species, neoplasms occur rarely or are not reported in invertebrates; therefore, they are of limited clinical importance to the veterinary practitioner. Invertebrates have been used in oncologic research for several decades, however, and as such, a few neoplasms are

Table 1
Examples of infectious agents causing systemic diseases or external lesions in various invertebrate species that can be diagnosed by cytology

| Etiology | Pathologic change | Host species | Additional information | Reference |
|---|---|---|---|---|
| **Cephalopods** | | | | |
| *Ichthyobodo necator* | 4–20-μm protozoa on skin surface and gill. Larger morph (15–20 μm) when on the gills and smaller size (4–8 μm) externally on the epidermis | Several octopus species | | [51] |
| **Bivalve mollusks** | | | | |
| Papovaviridae (virus-like inclusion) | Inclusions in hemolymph, gill epithelium, and connective tissue | Soft-shell clam (*Mya arenaria*) | | [19] |
| *Vibrio* sp | Gram-negative rods in hemolymph | Multiple bivalve mollusks | | [19,29] |
| *Marteilia refringens*, *Marteilia sydneyi* | Increased number of granulocytes, 5–8-μm (up to 40 μm during sporulation) sporoblasts surrounded by a halo in impression smear from the digestive gland | Several bivalve species (oyster, mussels) | Reportable disease | [26,29,49] |
| *Perkinsus marinus* | Protozoa (spherical bodies containing an eccentric vacuole "signet ring") in hemolymph, gills, and palp imprints | Several bivalve mollusk species | Reportable disease | [19,29] |
| *Bonamia ostreae* | Tissue imprint from gill and heart, extra and intracellular 1–5- μm spherical or ovoid protozoa with a central nucleus in hemocytes | Oyster species of the genus *Ostrea* | Reportable disease | [29,49,50] |
| Quahog parasite unknown (QPX), thraustochytrid-like disease of clams | Various stages of vegetative and nonstaining spore-like stages (2–25 μm in diameter) in abscesses or necrotic lesions, mantle, gills, and gonad (also in other organs) | American hard-shell clam (*Mercenaria mercenaria*) | | [30] |
| *Mikrocytos mackini* | Touch imprint from incised pustule from organs 2–3 μm in diameter, intra- or extracellular protozoa | Several species of oyster | Reportable disease | [30] |

| Organism | Cytologic findings | Host | Comments | References |
|---|---|---|---|---|
| *Giardia* sp, *Cryptosporidium* sp | None | Several shellfish species | Human pathogen, non pathogenic to shellfish | [28] |
| Crustaceans | | | | |
| Reo-like virus | Inclusion in hemocytes, multinucleated hemocytes | Various species of crab | Decreased clotting | [2,30] |
| Bifacies virus (Herpes-like virus disease) | Nuclear swelling, refractive intracytoplasmic inclusion | Various species of crab | Decreased clotting, increased hemolymph, turbidity in blue crab (*Callinectes sapidus*) | [21,29] |
| Hepatopancreatic parvovirus | Small eosinophilic inclusions or large basophilic inclusions within a hypertrophied nucleus on hepatopancreas impression smear | Prawn (*Penaeus* sp) | | [29] |
| *Baculovirus penaei* (BP virus disease) | Intranuclear occlusion bodies (0.1 to nearly 20 μm) in the epithelial cells of the hepatopancreas and midgut | Prawn (*Penaeus* sp) | Reportable disease | [29] |
| White spot baculovirus | Intranuclear lightly eosinophilic to deeply basophilic inclusions in hemocytes, epithelia, and connective tissues of the gills or stomach | Prawn (*Penaeus* sp) | | [29,47] |
| Yellowhead virus of *Penaeus monodon* | Abnormal hemocytes and connective tissue cells, with pyknotic and karyorrhectic nuclei and basophilic cytoplasmic inclusions | Giant tiger prawn (*Penaeus monodon*) | Reportable disease | [29] |
| *Vibrio* sp, *Aeromonas* sp | Gram-negative rods, hemocyte aggregates | Various aquatic invertebrate species | Decreased hemocyte counts and clotting time, pale, turbid hemolymph | [21,29] |
| *Aerococcus viridans* var *homari* (gaffkemia) | Tetrads of gram-positive cocci | Lobster (*Homarus* sp, *Panulirus* sp) | Decreased hemocyte counts and clotting time, pink and less viscous hemolymph | [2,13,29] |

(*continued on next page*)

Table 1 (continued)

| Etiology | Pathologic change | Host species | Additional information | Reference |
|---|---|---|---|---|
| Fusarium sp | Hyphae in cuticle, gills, and hemocoel | Marine and freshwater crustaceans | Increased hemocyte counts | [2,47] |
| Mesanophrys pugettmsis, Anophyroides haemophila (bumper car disease) | Amoeba | Various species of crab | Decreased hemocyte counts and clotting time, grey turbid hemolymph | [2,29,46] |
| Hematodinium sp | Nonmotile uni- and binucleate plasmodial parasite similar in size to hemocyte (5–14 μm in diameter) | Various species of crab, Norway lobster (Nephrops norvegicus) | Decrease hemocyte counts and clotting, turbid hemolymph | [2,21,30] |
| Paramoeba perniciosa | Amoeba in connective tissue and, in terminal stage, in hemolymph | Various species of crab, American lobster (Homarus americanus) | Cloudy hemolymph, decreased clotting | [2,29] |
| Microsporidian | Spores in imprint of affected muscle | Many species of crustacean | | [2,21,47] |
| Haplosporidian | Uninucleate and plasmodial stages of haplosporidian parasite in tissues | Many species of crustacean and shellfish | Opaque white hemolymph, decreased viscosity | [21,29] |

| | | | |
|---|---|---|---|
| *Microphallus basodactylophallus* (fluke) parasitized by *Urosporidium crescens* (haplosporidian) | Fluke cysts in thoracic muscles, hepatopancreas, and ventral ganglion | Many species of crab | Large brownish black metacercaria in tissue (called "pepper spot" or "buckshot") | [21,29] |
| Spiders | | | | |
| Nematodes (Panagrolaimidae family) | White discharge between mouth and chelicerae containing motile nematodes 0.5–3 mm in length | Theraphosidae spiders (tarantulas) | | [56] |
| Insects | | | | |
| Nuclear polyhedrosis viruses (Baculoviridae) | 0.5–15-μm nonstaining (Giemsa stain) intranuclear inclusion body in skin and hemocytes | Silkworm; numerous arthropod species, mostly in the Lepidoptera order | Host-specific viruses | [26,45,52] |
| Cytoplasmic polyhedrosis viruses (Baculoviridae) | 0.5–15-μm deeply stained (Giemsa) intracytoplasmic inclusion body in gut cells and hemocytes | Numerous arthropod species, mostly in the Lepidoptera order | Wider host range than nuclear polyhedrosis viruses | [45,52] |
| *Beauvaria bassiana* (muscardine) | Fungal hyphae | Silkworm; More than 500 insect species | | [26] |
| *Verticillium lecani* | Fungal hyphae | Various species | | [25] |
| *Aspergillus* sp *Fusarium* sp | Fungal hyphae | Multiple species | Opportunistic colonizer superficial wound | [9,27] |
| *Nosema apis* | Microsporidia with binucleated spores in the cytoplasm of midgut cells | Honeybee (*Apis mellifera*) | Reportable disease | [46] |
| *Ophryocystis elektroscirrha* | Small brown-gold football-shaped spores on the body scales | Monarch butterfly (*Danaus plexippus*) | | [55] |

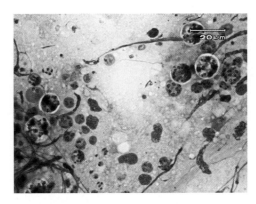

Fig. 4. Sporoblasts of *Marteilia refringens* from a digestive gland imprint from an oyster (*Ostrea eludis*). (*From* Levine JF, Law M, Corsin F. Bivalves In: Lewbart GA, editor. Invertebrate medicine. Ames (IA): Blackwell Publishing; 2006. p. 91–113, Color plate 7.12; with permission.)

well described in some invertebrate species [31,32]. The cellular neoplastic changes are comparable to those observed in mammals, and sampled cells should be evaluated for cytologic criteria of malignancy (ie, cellular pleomorphism, hypercellularity, lack of differentiation, rapid cell division); if sufficient criteria are present, a diagnosis of neoplasia may be made [11,23]. Definitive diagnosis of neoplasia by cytology is difficult, and submission of samples for histopathologic examination for confirmation of the diagnosis is advised. A classic example of neoplasia in invertebrates is the hematopoietic neoplasm (clam leukemia or disseminated sarcoma-neoplasia) in soft-shell clams that is characterized by the presence of circulating, large, anaplastic hemocytes that have hyperchromatic and often

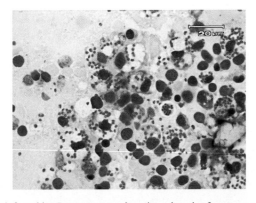

Fig. 5. Hemocytes infected by *Bonamia ostrea* in a tissue imprint from an oyster (*Ostrea eludis*). Numerous round basophilic protozoa 2 to 3 μm in diameter are present in the hemocytes. (*From* Levine JF, Law M, Corsin F. Bivalves In: Lewbart GA, editor. Invertebrate medicine. Ames (IA): Blackwell Publishing; 2006. p. 91–113, Color plate 7.14; with permission.)

pleomorphic nuclei with variably prominent nucleoli and small, intracytoplasmic, clear vacuoles that stain with oil red O [33]. Studies evaluating the correlation between in vivo cytologic diagnosis and histologic diagnosis are exceptional. One study on hematopoietic neoplasm in soft-shell clams (*Mya arenaria*) found the accuracy of the in vivo cytologic diagnosis to range from 66% to 100% depending on disease severity [11]. For a review of neoplasms reported in invertebrates, the reader is directed to an article by Peters [34]. Additional information can be obtained at the Registry of Tumor in Lower Animals [35] and in relevant published literature.

*Hemolymph evaluation*

In several invertebrate species, there is good evidence that changes occur in the hemolymph after infection with various diseases or toxic insults, because they occur in the blood of vertebrates [2,6,36]. The major obstacle to interpretation is the lack of normal physiologic reference values and variation observed with environmental factors, such as temperature, nutritional status, molt stage, and salinity for aquatic species, however [2]. The main parameters that can be evaluated cytologically are clotting time, hemolymph color, THC and differential hemocyte count (DHC), hemocyte morphology, and presence of infectious agents. Given the absence of or limited information on normal coagulation time, hemocyte counts for specific species, and wide variation engendered by the environment and physiologic status, the best results are obtained when samples from sick invertebrates are evaluated in conjunction with samples from healthy individuals of the same species maintained under similar conditions.

*Clotting time*

Hemolymph clotting is included in this article because this parameter is easily evaluated at the time of sample collection and represents important data when interpreting the hemolymph sample. Hemolymph clotting is an integral part of the invertebrate immune response and protects against loss of body fluid, prevents organism invasion, and sequesters invading organisms. Numerous diseases are reported to decrease clotting of hemolymph, but no detailed information or commercially available tests exist for any invertebrate species. Normal hemolymph generally coagulates readily after collection; therefore, slow clotting hemolymph is considered abnormal and indicative of a diseased state.

*Hemolymph color*

Hemolymph color is an easily visible marker of health. The hemolymph clarity and color can be variable but is usually colorless or various shades of pale blue because of the copper-containing respiratory pigment hemocyanin [2]. In disease states, changes in turbidity and color may be observed, with hemolymph most often becoming more turbid and white. A classic example of hemolymph changes associated with a specific disease is the pink hemolymph color of American lobsters (*Homarus americanus*) affected with gaffkemia (*Aerococcus viridans* var *homar*) [2,13]. The lack of color may be indicative of low hemocyanin content in species in which the hemolymph is not normally colorless [2].

*Hemocyte evaluation*
*Normal hemocyte identification.* In general, the terminology associated with the type of hemocytes is confusing, and there is no solid agreement on the classification of the different hemocyte types. The exact role of each cell type is also not well defined. There is variation in the number and type of hemocytes reported depending on taxa, and these parameters often vary for the same species depending on the authors [2,13,37–39]. Published information on hemocyte morphology in a few common taxa is presented here. The reader is directed for information on specific species to the published literature and the 15-volume treatise, *Microscopic Anatomy of Invertebrates* [40]. In crustaceans, in which significant study has been conducted, hemocytes are most often divided into three types: hyalinocytes (hyaline cell and hyaline hemocytes or amebocytes) and semigranular (small-granule) and granular (large-granule) cells (hemocytes and amebocytes as well as granulocytes) [2,13,41]. It is not clear if these represent different cells or are different stages of the same cells, however [2]. Granular cells are the largest, with prominent intracytoplasmic granules ($>1$ μm in diameter) and eccentric nuclei (Fig. 6). They are involved in phagocytosis and encapsulation of foreign bodies or large infectious agents, at least in shrimp, lobsters, and crabs. Their granules contain lysosomal enzymes, such as acid phosphatase, β-glucuronidase, and prophenoloxidase. The semigranular cells are intermediate in appearance between hyalinocytes and granular cells and are

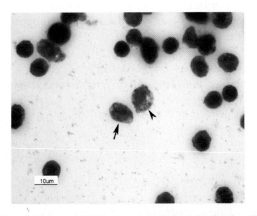

Fig. 6. Hemocytes from a blue crab (*Callinectes sapidus*), formalin fixation. The arrow indicates a semigranular hemocyte. The arrowhead indicates a granular hemocyte (Wright-Giemsa stain, original magnification ×100).

also involved in coagulation or phagocytosis, depending on the species [2]. Semigranular cells have fewer intracytoplasmic granules of variable size but are less than 1 μm in diameter and have a central nucleus. Hyalinocytes are the smallest and most variable type of hemocyte. They have a higher nuclear-to-cytoplasmic ratio than the semigranular and granular cells and have only a few small intracytoplasmic granules. They are involved in initiating coagulation, tanning of the cuticle, and, in some species (eg, shore crab [*C maenas*], crayfish [Cambaridae]), phagocytosis [5,6,42]. Bivalve mollusk hemocytes are generally separated into granulocytes and hyalinocytes, with some exceptions, such as the Pacific oyster (*Crassostrea gigas*), in which two types of granulocytes are recognized (granulocytes and intermediate cells), and scallops, in which granulocytes are absent [10]. Gastropods were considered to have only one type of hemocyte, but more recent work suggests that, at least in the snail *Biomphalaria glabrata*, there are three distinct hemocyte types. These observations were made using electron microscopy, however, and differences are likely difficult to identify using light microscopy [43]. Cephalopods are reported to have only one type of hemocyte that does not have phagocytizing capacity [44]. Spiders are reported to have hemocytes with dense cytoplasmic granules (granulocytes), hemocytes with a single secretory vacuole (leberiridiocytes), and cyanocytes that contain hemocyanin in a crystalline form before secretion (Figs. 7 and 8) [4,8]. Insect hemocytes have been described in species from varied orders, and the most common types of hemocytes reported include prohemocytes, granulocytes (granular cells), plasmatocytes, spherulocytes (spherule cells), and oenocytoids. The morphology of these cells varies among insect orders, however, and the classification scheme used by the different authors is inconsistent [37,45]. Granular cells are distinguished by abundant cytoplasmic granules and are involved in coagulation. Plasmatocytes are polymorphic

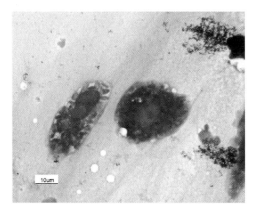

Fig. 7. Cyanocytes from a rosehair tarantula (*Grammostola rosea*). Bouin fixation (Wright-Giemsa stain, original magnification ×100).

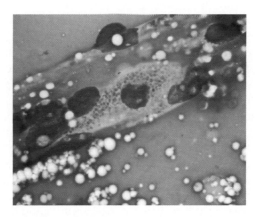

Fig. 8. Granular hemocyte from a rosehair tarantula (*Grammostola rosea*). No fixative solution (Wright-Giemsa stain, original magnification ×100).

cells that are oval or spindle shaped, may contain few intracytoplasmic granules, and are involved in phagocytosis and encapsulation. Oenocytoids are large spherical cells with a small central nucleus; their function is unknown. Prohemocytes are small round cells with a large nucleus and are considered stem cells [37,38]. *Drosophila* larvae have three circulating hemocyte types that are named somewhat differently from those of most other insects: plasmatocytes, lamellocytes, and crystal cells [37]. It is accepted that granular cells and plasmatocytes are the norm in Lepidoptera, whereas in *Drosophila* larvae, plasmatocytes and lamellocytes are the cells predominantly involved in the cellular response, with the latter being specifically devoted to encapsulation [37,38].

*Hemocyte morphologic alteration.* Given that cell morphology is often not well described, degranulation occurs easily, and hemolymph anticoagulant and preservation techniques are not extremely efficient and standardized, no specific reliable morphologic changes can be used in a clinical situation, with the possible exception of mitotic figures in crustaceans. Hemocyte mitoses are normally restricted to the hematopoietic tissue, and observation of hemocyte mitotic figures in a hemolymph sample is suggestive of an inflammatory reaction [13].

*Hemocyte count.* Use of the THC and DHC has been reported to evaluate hemolymph samples of crustaceans and bivalve mollusks [13,20]. Flow cytometry is used successfully in some invertebrate species, but this method is not available to most clinicians and is not discussed further. Direct manual counting is performed using a hemocytometer. This technique is time-consuming but is amenable to clinical practice. The counting method is similar to that used in vertebrate medicine, and the hemolymph or formalin-diluted hemolymph may be mixed directly with the staining solution

Table 2
Examples of hemocyte counts

| Host species | Total hemocyte count | Reference |
| --- | --- | --- |
| American lobster (*Homarus americanus*) | $8.1–22.4 \times 10^6$ cells/mL | [13] |
| Blue crab (*Callinectes sapidus*) | $4–128 \times 10^6$ cells/mL | [57] |
| Giant clam (*Tridacna crocea*) | $0.3–2.6 \; 10^6$ cells/mL | [36] |
| Blackfoot abalone (*Haliotis iris*) | $8.8–11.6 \times 10^6$ cells/mL | [53] |
| Eastern elliptio (*Elliptio complanata*) | $2.50–23 \times 10^6$ cells/mL | [54] |

[22,36]. There is variation in hemocyte counts between the different species that have been evaluated, and data on normal hemocyte counts are sparse. Examples are given in Table 2. When available, the easiest method is to collect a sample from healthy specimens, housed in the same environment if possible, for comparison. Generally, decreased circulating hemocytes are observed in many infectious and noninfectious disease states but may also be observed near molting [2]. Stress, such as bleeding and emersion, was reported to increase hemocyte counts within 1 hour [2]. In bivalve mollusks, increased hemocyte counts were observed in experimental clams and oysters after bacterial or chemical contamination [10,35]. For the clinician or clinical pathologist involved in exotic animal practice, the DHC may be difficult to perform, given the wide variety of invertebrates encountered and the variation in cell morphology and classification schemes used in different species. An easy and practical method is to separate the hemocytes observed into granular and agranular hemocytes, because most authors agree that granular and agranular hemocytes can be readily recognized and that granular hemocytes and hyalinocytes are most prevalent and predominantly involved in the cellular response [2,13,35,37]. The differential hemocyte count is difficult to interpret, because wide species and interanimal variations are present [13]. Nevertheless, information about variation in the DHC is available for some species. As an example, in oysters, it has been shown that agranular hemocytes predominate in response to the protozoan parasite *Haplosporidium nelsoni*, whereas granular cells are the most responsive to bacteria and viruses [35]. Other hemolymph parameters can be evaluated, such as hemolymph endobiotics, hemocyanin, total protein, osmolality, electrolytes, acid-base balance, hormones, and glucose, but this is outside the scope of this article.

**Summary**

Invertebrate medicine in the context of an exotic or zoo animal veterinary practice is in its infancy. Establishment of species-specific reference values and evaluation of the effectiveness of cytology for diagnosis of specific diseases are necessary. Despite the lack of normal reference parameters for most species encountered in clinical practice, important information may be obtained from cytologic examination of tissue imprints, aspirates, scrapings, and hemolymph. This information may be essential to establish a specific diagnosis, focus investigations, and influence treatments. It is hoped that this article stimulates veterinarians who work with invertebrates to use diagnostic cytology and disseminate the results of their experience [48].

## Acknowledgments

The authors thank Dr. E. Noga for his expert advice, S. Davis for sharing her experience, and Dr. J. Tarigo and L. Christian for help with the images.

## References

[1] Ruppert EE, Barnes RD. Invertebrate zoology. 6th edition. Orlando (FL): Harcourt Brace College Publishers; 1994. p. 928.

[2] Noga EJ. Hemolymph biomarkers of crustacean health. In: Fingerman M, Nagabhushanam R, editors. Recent advances in marine biotechnology, immunobiology and pathology, vol. 5. Enfield (NH): Science Publishers; 2000. p. 124–63.

[3] Ruppert EE, Fox RS, Barnes RD. Invertebrate zoology: a functional evolutionary approach. 7th edition. Belmont (CA): Thompson-Brooks/Cole; 2004. p. 963.

[4] Foelix RF. Biology of spiders. New York: Oxford University Press; 1996. p. 336.

[5] Goins KR. Host defense mechanisms in the crayfish: the effect of injection with live or killed bacteria [master's thesis]. Johnson City (TN): East Tennessee State University; 2003. p. 41.

[6] Iwanaga S, Lee BL. Recent advances in the innate immunity of invertebrate animals. J Biochem Mol Biol 2005;38(2):128–50.

[7] Robertson M, Postlethwait JH. The humoral antibacterial response of *Drosophila* adults. Dev Comp Immunol 1986;10:167–79.

[8] Pizzi R. Spiders. In: Lewbart GA, editor. Invertebrate medicine. Ames (IA): Blackwell Publishing; 2006. p. 143–68.

[9] Gunkel C, Lewbart GA. Invertebrate anesthesia. In: Heard D, editor. Zoo animal and wildlife immobilization and anesthesia. Ames (IA): Blackwell Publishing; in press.

[10] Auffret M, Oubella R. Cytological and cytometric analysis of bivalve mollusc hemocytes. In: Stolen JS, Fletcher TC, Smith JT, et al, editors. Techniques in fish immunology, vol. 4. Immunology and pathology of aquatic invertebrates. 1st edition. Fair Haven (NJ): SOS Publication; 1995. p. 55–64.

[11] Cooper KR, Brown RS, Chang PW. Accuracy of blood cytological screening techniques for the diagnosis of a possible hematopoietic neoplasm in the bivalve mollusk, *Mya arenaria*. J Invert Pathol 1982;39(3):281–9.

[12] Santarem M, Figueras A. Basic studies on defense mechanism of mussels. In: Stolen JS, Fletcher TC, Smith JT, et al, editors. Techniques in fish immunology, vol. 4. Immunology and pathology of aquatic invertebrates. 1st edition. Fair Haven (NJ): SOS Publication; 1995. p. 87–92.

[13] Battison A, Cawthorn R, Horney B. Classification of *Homarus americanus* hemocytes and the use of differential hemocytes counts in lobsters infected with *Aerococcus viridans* var. *homari* (gaffkemia). J Invert Pathol 2003;84:177–97.

[14] Cooper JE. Bleeding of pulmonate snails. Lab Anim 1994;28:277–8.

[15] Smith SA. Horseshoe crabs. In: Lewbart GA, editor. Invertebrate medicine. Ames (IA): Blackwell Publishing; 2006. p. 133–42.

[16] Lewbart GA, editor. Invertebrate medicine. Ames (IA): Blackwell Publishing; 2006. p. 327.

[17] Cooper JE. Emergency care of invertebrates. Vet Clin North Am Exot Anim Pract 1998;1(1): 251–64.
[18] Hurton L, Berkson J, Smith S. Estimation of total hemolymph volume in the horseshoe crab *Limulus polyphemus*. Marine and Freshwater Behaviour and Physiology 2005;38(2):139–47.
[19] McLaughlin SM. Diagnostic techniques for the softshell clam *Mya arenaria*, Linnaeus, 1758. In: Howard DW, Lewis EJ, Keller BJ, et al, editors. Histological techniques for marine bivalve mollusks and crustaceans. Oxford (MD): NOAA Technical Memorandum NOS NCCOS 5. 2004. p. 169–78.
[20] Allam B, Ashton-Alcox KA, Ford SE. Haemocyte parameters associated with resistance to brown ring disease in *Ruditapes* spp. clams. Dev Comp Immunol 2001;25:365–75.
[21] Messick GA. Laboratory techniques for detecting parasites and diseases of blue crabs, *Callinectes sapidus*. In: Stolen JS, Fletcher TC, Smith JT, et al, editors. Techniques in fish immunology, vol. 4. Immunology and pathology of aquatic invertebrates. 1st edition. Fair Haven (NJ): SOS Publication; 1995. p. 187–99.
[22] Messick GA. Diagnostic techniques for blue crabs. In: Howard DW, Lewis EJ, Keller BJ, et al, editors. Histological techniques for marine bivalve mollusks and crustaceans. Oxford (MD): NOAA Technical Memorandum NOS NCCOS 5. 2004. p. 151–7.
[23] Tyler RD, Cowell RL, Baldwin CJ, et al. Introduction. In: Cowell RL, Tyler RD, Meinkoth JH, editors. Diagnostic cytology and hematology of the dog and cat. 2nd edition. St. Louis (MD): Mosby; 1999. p. 1–19.
[24] Cooper JE, Cunningham AA. Pathological investigation of captive invertebrates. International Zoo Yearbook 1991;30:137–43.
[25] Cunningham AA, Frank JM, Croft P, et al. Mortality of captive British wartbiter crickets: implications for reintroduction programs. J Wildl Dis 1997;33(3):673–6.
[26] Williams DL. Sample taking in invertebrate veterinary medicine. Vet Clin North Am Exot Anim Pract 1999;2(3):777–801.
[27] Williams DL. Invertebrates. In: Meredith A, Redrobe S, editors. Manual of exotic pets. 4th edition. Gloucester (UK): British Small Animal Veterinary Association; 2002. p. 280–7.
[28] Lewis EJ. Miscellaneous techniques and methods for mollusks. In: Howard DW, Lewis EJ, Keller BJ, et al, editors. Histological techniques for marine bivalve mollusks and crustaceans. Oxford (MD): NOAA Technical Memorandum NOS NCCOS 5. 2004. p. 159–68.
[29] Bower S, McGladdery SE, Price IM. Synopsis of infectious diseases and parasites of commercially exploited shellfish. Annual Review of Fish Diseases 1994;4:1–199.
[30] Bower S, McGladdery SE, Price IM. Synopsis of infectious diseases and parasites of commercially exploited shellfish. Available at: http://www.pac.dfo-mpo.gc.ca/sci/shelldis/title_e.htm. Accessed August 28, 2006.
[31] Cooper JE. Oncology of invertebrates. Vet Clin North Am Exot Anim Pract 2004;7(3): 697–703.
[32] Vidal M, Cagan RL. *Drosophila* models for cancer research. Curr Opin Genet Dev 2006; 16(1):10–6.
[33] Kelley ML, Winge P, Heaney JD, et al. Expression of homologues for p53 and p73 in the softshell clam (*Mya arenaria*), a naturally-occurring model for human cancer. Oncogene 2001;20:748–58.
[34] Peters EC. Invertebrate neoplasms. In: Lewbart GA, editor. Invertebrate medicine. Ames (IA): Blackwell Publishing; 2006. p. 297–9.
[35] Registry of Tumor in Lower Animals. Available at: http://www.pathology-registry.org/index_1.asp. Accessed November 28, 2006.
[36] McCormick-Rayl MG, Howard T. Morphology and mobility of oyster hemocytes: evidence for seasonal variations. In: Proceeding of the International Association for Aquatic Animal Medicine. Marineland (FL); 1991. p. 145–8.
[37] Nakayama K, Nomoto AM, Nishijima M, et al. Morphological and functional characterization of hemocytes in the giant clam *Tridacna crocea*. J Invert Pathol 1997;69:105–11.

[38] Lavine MD, Strand MR. Insect hemocytes and their role in immunity. Insect Biochem Mol Biol 2002;32:1295–309.
[39] Willott E, Trenczek T, Thrower LW, et al. Immunochemical identification of insect hemocytes populations: monoclonal antibodies distinguish four major hemocytes types in Manduca Sexta. Eur J Cell Biol 1994;65:417–23.
[40] Harrison FW, editor. Microscopic anatomy of invertebrates, 15 vols. New York: Wiley-Liss; 1991–1997.
[41] Sis RF, Lewis DH, Means JE. Shrimp: an anatomical study. In: Proceedings of the International Association for Aquatic Animal Medicine. Hong Kong: 1992. p. 52–60.
[42] Bell KL, Smith V. In vitro superoxide production by hyaline cells of the shore crab *Carcinus maenas*. Dev Comp Immunol 1993;17:215–23.
[43] Matricon-Gondran M, Letocart M. Internal defenses of the snail *Biomphalaria glabrata*. I. Characterization of hemocytes and fixed phagocytes. J Invert Pathol 1999;74:224–34.
[44] Oestmann DJ, Scimeca JM, Forsythe J, et al. Special consideration for keeping cephalopods in laboratory facilities. Contemp Top Lab Anim Sci 1997;36(2):89–93.
[45] Boucias DG, Pendland JC. Principles of insect pathology. Boston: Kluwer Academic Publisher; 1998. p. 537.
[46] Cawthorn RJ. Overview of "bumper car" disease—impact on the North American lobster fishery. Int J Parasitol 1997;27(2):167–72.
[47] Edgerton BF, Evans LH, Stephens FJ, et al. Synopsis of freshwater crayfish diseases and commensal organisms. Aquacult 2002;206:57–135.
[48] Noga EJ, Hancock AL, Bullis RA. Crustaceans. In: Lewbart GA, editor. Invertebrate medicine. Ames (IA): Blackwell Publishing; 2006. p. 179–94.
[49] da Silva PM, Villalba A. Comparison of light microscopic techniques for the diagnosis of the infection of the European flat oyster *Ostrea edulis* by the protozoan *Bonamia ostreae*. J Invert Pathol 2004;85:97–104.
[50] Levine JF, Law M, Corsin F. Bivalves. In: Lewbart GA, editor. Invertebrate medicine. Ames (IA): Blackwell Publishing; 2006. p. 91–113.
[51] Forsythe JW, Hanlon RT, Lee PG. A synopsis of cephalopod pathology in captivity. In: Proceedings of the International Association for Aquatic Animal Medicine. San Antonio (TX): 1989. p. 109–15.
[52] Rivers CF. The control of diseases in insect cultures. International Zoo Yearbook 1991;30: 131–7.
[53] Smolowitz R. Gastropods. In: Lewbart GA, editor. Invertebrate medicine. Ames (IA): Blackwell Publishing; 2006. p. 65–78.
[54] Gustafson LL, Stoskopf MK, Showers W, et al. Reference ranges for hemolymph chemistries from *Elliptio complanata* of North Carolina. Dis Aquat Organ 2005;65:167–76.
[55] Cooper JE. Insects. In: Lewbart GA, editor. Invertebrate medicine. Ames (IA): Blackwell Publishing; 2006. p. 205–19.
[56] Pizzi R, Carta L, George S. Oral nematode infection of tarantulas. Vet Rec 2003;52:695.
[57] Sawyer TK, Cox R, Higginbottom M. Hemocyte values in healthy blue crabs, *Callinectes sapidus*, and crabs infected with the amoeba, *Paramoeba perniciosa*. J Invert Pathol 1970; 15:440–6.

# VETERINARY CLINICS
Exotic Animal Practice

# Index

*Note:* Page numbers of article titles are in **boldface** type.

## A

Abdominal effusion, cytologic diagnosis of, in ferrets, 74–75

Abscesses, cytologic diagnosis of, in prairie dogs, facial, 48
in rabbits, 33–34

Actinomycosis, cytologic diagnosis of, in ferrets, 70

Adenocarcinoma, cytologic diagnosis of, mammary gland, in hamsters, 46
in hedgehogs, 52–53
uterine, in rabbits, 29–30

Adenoma, cytologic diagnosis of, in guinea pigs, 37
in hamsters, 46

Adipocyte tumors, cytologic diagnosis of, in guinea pigs, 35–36

Adrenal cortical adenoma, cytologic diagnosis of, in hamsters, 46

Adrenocortical carcinoma, cytologic diagnosis of, in hedgehogs, 54–55

*Aeromonas* complex, cytologic diagnosis of, in fish, 215

Air sac imprint lesions, cytologic diagnosis of, in avians, 143

Algae, cytologic diagnosis of, in marine mammal nasal sac, 225

Algae-induced dermatitis, cytologic diagnosis of, in fish, 217–218

Amoebae, cytologic diagnosis of, in invertebrates, 244

Amphibians, cytologic diagnosis in, **187–206**
methods for, 187
of cloacal wash, 200
of coelomic effusions, 199–200
of feces, 200
of infections, 189–203
of lymph sac, subcutaneous, 199–200
of neoplasia, 203–204
of noninfectious diseases, 203–204
of peripheral blood, 187–190
of skin, 190–199
of visceral organs, 200–203
collection techniques for, 200–201
normal features of, 201–202
integument of, chromatophore hyperplasia of, 198–199
clinical sample techniques for, 191–192
functions of, 190–191
infectious diseases of, 192–198

Amyloidosis, cytologic diagnosis of, in avians, 150–151

Anisokaryosis, 17, 164–165

Anticoagulants, for hemolymph smear preparation, 238–239, 250

Apocrine gland tumors, cytologic diagnosis of, in ferrets, 67–68

*Argulus* sp., cytologic diagnosis of, in fish, 221–222

Arthropods, hemolymph collection in, 237
infectious disease cytology in, 245

Artifacts, in cytology specimens, 3, 21, 76
of amphibians, 189–190
of avians, 151–152
of marine mammals, 97–99, 103, 105–107, 112–113, 115, 125

Ascitic fluid, cytologic diagnosis of, in fish, 210–211

Aspirates, for cytologic diagnosis, in avians, 137–140, 147–148
in ferrets, 74, 76
in fish, 208, 230
in hedgehogs, 53–55, 57–59
in marine mammals, 127
in rabbits, 26–31, 34, 36, 38–39, 45, 47
needle. See *Fine-needle aspirates (FNAs)*.

Aspiration pneumonia, cytologic diagnosis of, in ferrets, 76

*Atoxoplasma* sp., cytologic diagnosis of, in avians, 149–150

Aviadenovirus infection, cytologic diagnosis of, in avians, 150–151

Avians, cytologic diagnosis in, **131–154**
    common artifacts with, 151–152
    of infection, 134–137, 140–141, 147, 149
        gastrointestinal, 143–146
        respiratory, 141–142
    of inflammation, 132–134, 140
    of neoplasia, 135–136
    of specific anatomic sites, 136–151
        air sac imprints, 143
        body cavity effusions, 146–147
        choanal slit, 143–146
        cloaca, 145–146
        conjunctiva, 140–141
        crop, 143–146
        emerging feather, 136–140
        esophagus, 143–146
        feather follicles, 136–140
        gastrointestinal tract, 143–146
        infraorbital sinuses, 141–143
        liver, 147–151
        lung, 143
        nasal sinuses, 141–143
        oral cavity, 143–146
        respiratory tract, 141–143
            lower, 143
            upper, 141–143
        skin, 136–140
        spleen, 147, 151
        subcutis, 136–140
        synovial fluid, 147–148
        trachea, 141–143
        vent, 145–146
    problem areas of, 152–153
    specimen techniques for, 131
    stains for, 131

# B

Bacterial infections, cytologic diagnosis of, 3, 12
    in amphibians, 192–193, 195, 202
    in avians, 134–136, 141, 147
        gastrointestinal, 143–146
        respiratory, 141–143
    in ferrets, 62–63, 70–71
    in fish, 210, 213–217, 223–225, 227, 229
    in hedgehogs, 56–57
    in invertebrates, 241–243
    in marine mammals, 81–82, 85, 89
        gastrointestinal, 102–103, 105–111, 113
        opportunistic respiratory, 92–94, 96–97
    in mice, 44
    in reptiles, 184

Basal cell tumors, cytologic diagnosis of, in ferrets, 66–67
    in rabbits, 25–26

Basophils, cytologic diagnosis of, in amphibians, 188–189
    in avians, 133–134
    in hemic tissue, 8–9
    in reptiles, 160, 167

*Batrachochytrium dendrobatidis,* cytologic diagnosis of, in amphibians, 195–196

Benign neoplasia, cytologic diagnosis of, 15–16. See also *specific pathology or tumor, e.g.,* Squamous papilloma.
    in amphibians, 203–204
    in avians, 135–136
    in ferrets, 63–65
    in reptiles, 163–168
    species indications for. See *specific species.*

*Blastomyces dermatitidis,* cytologic diagnosis of, in ferrets, 71–72

Blood contamination, of cytologic specimens, 3, 21
    in reptiles, 162, 177

Blowhole, cytologic diagnosis of, in marine mammals, 89–95
        sampling technique for, 87–88

Body cavity effusions, cytologic diagnosis of, in avians, 146–147
    in ferrets, 74–75
    in marine mammals, 127
    in reptiles, 169–175
        approach to, 169–170
        coelomic classifications of, 170
        exudates, septic vs. nonseptic, 170, 172–173
        hemorrhagic, 170, 173–175
        melanocytes in, 171
        mesothelial cells in, 169–171
        modified transudates, 170–172

neoplastic, 170, 175
transudates, 170–171
*Bonamia ostrea,* cytologic diagnosis of, in invertebrates, 242, 246
Branchitis, cytologic diagnosis of, in fish, 228
Bronchoalveolar lavage/brushings, cytologic diagnosis of, in marine mammals, 92, 95
Bronchogenic papillary adenoma, cytologic diagnosis of, in guinea pigs, 37
Bronchopneumonia, cytologic diagnosis of, in rats, 42
*Brucella* sp., cytologic diagnosis of, in marine mammals, 81–82
Buffer recipes, for hemolymph smear preparation, 239
Butterfly, hemolymph collection in, 236
infectious disease cytology in, 245

## C

Calcium deposits, in synovial fluid, of reptiles, 177–179
*Camallanus* sp., cytologic diagnosis of, in fish, 211–212
*Candida* sp., cytologic diagnosis of, in avians, 144
in marine mammals, 85–86, 92
*Capillaria* sp., cytologic diagnosis of, in avians, 144–145
Carcinomas, cytologic diagnosis of. See also *specific pathology, e.g.,* Squamous cell carcinoma.
epithelial, 17
in avians, 147
in reptiles, 175
Cardiomyopathy, cytologic diagnosis of, in hedgehogs, 58–59
Casts, urinary, in marine mammals, 118–120
Cell nucleus, in cytologic diagnosis, characteristics of, 2
Cellular appearance, in cytologic diagnosis, importance of, 2
Centrifugation, of marine mammals samples, gastric, 98, 101, 105
urinalysis, 116
Cephalopods, hemocyte identification in, 249–250
infectious disease cytology in, 242

Cerebrospinal fluid (CSF), cytologic diagnosis of, in marine mammals, 127
Cervical lymphadenitis, cytologic diagnosis of, in guinea pigs, 39
Cetaceans, cytologic diagnosis in, **79–130**
literature review of, 81
of feces, 106–111
of gastrointestinal tract, 97–105
of infections, 81–86
of inflammatory lesions, 83–86, 90–91, 102–103, 110–111, 117
of milk, 123–127
of miscellaneous specimens, 127
of respiratory tract, 86–97
of urine, 113–120
of vagina, 121–123
pathologic significance evaluation, 83–86
public health considerations, 81–82
restraint for sample collection, 82–83
species overview, 79–81, 127
upper respiratory tract uniqueness of, 98–99
Chemotactic factors, of inflammation exudates, 10
Chinchillas, cytologic diagnosis in, of nonneoplastic conditions, 48
Chlamydophilosis, cytologic diagnosis of, in amphibians, 202
in avians, 141
Choanal slit lesions, cytologic diagnosis of, in avians, 143–146
Cholesterol-to-triglyceride ratio, in pseudochylous effusions, 22
Chordoma, cytologic diagnosis of, in ferrets, 69–70
Chromatia, in malignant neoplasms, 164–165
Chromatin patterns, nuclear, in cytologic diagnosis, 2
Chromatophore hyperplasia, cytologic diagnosis of, in amphibians, 198–199
in fish, 222–223
Chromomycosis, cytologic diagnosis of, in amphibians, 193–194
Chylous effusions, cytologic diagnosis of, 22
Chytridiomycosis, cytologic diagnosis of, in amphibians, 195–196

Clams, hard shell. See *Mollusks*.
  soft-shell, cytologic diagnosis in,
    hemocyte count, 251
    of infections, 242
    of neoplasia, 246–247
Cloacal lesions, cytologic diagnosis of, in
    amphibians, 200
  in avians, 145–146
Cloacal swabs, for cytologic diagnosis, in
    reptiles, 181, 184–185
*Clostridium* sp., cytologic diagnosis of, in
    avians, 145–146
  in marine mammals, 81–82
Clotting time, of hemolymph, 247
Coccidial infections, cytologic diagnosis of,
    in fish, 213, 227
  in marine mammals, 114, 125
Coelomic effusions, cytologic diagnosis of,
    in amphibians, 199–200
  in fish, 210–211
  in reptiles, 170
Coelomitis, egg yolk, cytologic diagnosis of,
    in avians, 146–147
Conjunctival lesions, cytologic diagnosis of,
    in avians, 140–141
Connective tissue lesions, cytologic
    diagnosis of, 3–4
  in ferrets, 63–65, 69–70
Crab, hemocyte identification in, 248–249
  hemolymph collection,
    237–238
  infectious disease cytology in,
    243–244
Crayfish, hemocyte identification in, 249
  hemolymph collection in, 238
Crop lesions, cytologic diagnosis of, in
    avians, 143–146
Crustaceans, hemocyte identification in, 248
  infectious disease cytology in, 243–244
*Cryptococcus neoformans,* cytologic
    diagnosis of, in avians, 142
Cryptosporidiosis, cytologic diagnosis of, in
    fish, 212–213
  in reptiles, 183–185
Crystals, cytologic diagnosis of, gouty. See
    *Uric acid crystals.*
  hemosiderin. See *Hemosiderin
    crystals.*
  in marine mammals, salt, 99
    urinary, 118–120

Cystic lesions, cytologic diagnosis of, in fish,
    224–225
  in guinea pigs, ovarian, 40
  in reptiles, 160–162
Cystocentesis, for cytologic samples, in
    marine mammals, 115
Cytologic diagnosis, 1. See also *specific
    species.*
  basics of, **1–24**. See also *Cytology.*
  in amphibians, **187–206**
  in avians, **131–154**
  in cetaceans and sirenians, **79–130**
  in ferrets, **61–78**
  in fish, **207–234**
  in hedgehogs, **51–59**
  in invertebrates, **235–254**
  in rabbits, guinea pigs, and rodents,
    **25–49**
  in reptiles, **155–186**
Cytology, basics of, **1–24**
  cell nucleus characteristics in, 2
  cellular appearance importance
    to, 2
  common artifacts found in,
    151–152, 189
  connective tissue in, 3–4
  cytoplasmic volume in, 2
  diagnostic role of, 1. See also
    *specific species.*
    in amphibians, **187–206**
    in avians, **131–154**
    in cetaceans and sirenians,
      **79–130**
    in ferrets, **61–78**
    in fish, **207–234**
    in hedgehogs, **51–59**
    in invertebrates, **235–254**
    in rabbits, guinea pigs, and
      rodents, **25–49**
    in reptiles, **155–186**
  effusions in, 19–22. See also
    *Effusions.*
    chylous, 22
    examination components
      for, 19
    exudative, 21
    hemorrhagic, 21–22
    malignant (neoplastic), 22
    mesothelial cell
      proliferation and,
      20–21
    pseudochylous, 22
    transudative, 19–20
  epithelial-glandular tissue in, 3
  hemic tissue in, 3–9
    basophils as, 8–9
    eosinophils as, 7–8
    erythrocytes as, 4–5

heterophils as, 6–7
importance of, 3–4
leukocytes as, 4–5
lymphocytes as, 5
monocytes as, 5
neutrophils as, 5–6
platelets as, 9
thrombocytes as, 9
inflammation exudates in, 3, 10–15
  chemotactic factors of, 10
  eosinophilic, 15
  granulocyte degeneration and, 11–12
  heterophilic, 10–11
    classification of, 11–12
    process of, 12
  histiocytic, 13–14
  lymphocytic, 12–14
  macrophagic, 12–14
  melanomacrophages and, 14
  mixed cell, 12–13
  neutrophilic, 10–11
    classification of, 11–12
    process of, 12
  nuclear pyknosis and, 12
  phagocytic, 10–11, 14
  plasmacytic, 12, 14–15
  process stages of, 10
  septic, 12
neoplasia in, benign, 15–16
  malignant, 16–19
    cytoplasmic cellular features of, 16–17
    diagnostic categories for, 16
    epithelial cell origins, 17
    general cellular features of, 16
    mesenchymal origins, 17–18
    nuclear features of, 16–17
    round cell origins, 18–19
    structural features of, 16–19
nervous tissue in, 3
noncellular characteristics in, 2–3
quality impressions in, 2
scanning magnification for, 2
specimen collection techniques for. See *specific species.*
stains used in, 1–2
tissue classifications for, 3
tissue hyperplasia in, 15–16

Cytoplasm, neoplastic features of, in reptiles, 166–167

Cytoplasmic inclusions, cytologic diagnosis of, in amphibians, 189–190

Cytoplasmic volume, in cytologic diagnosis, 2

## D

*Dactylogyrus* sp., cytologic diagnosis of, in fish, 220–221

Dermal fibrosis, nodular, in rabbits, 32–33

Dermatitis, algae-induced, in fish, 217–218

Dermatophytosis, cytologic diagnosis of, in ferrets, 70–71

Diatoms, in marine mammal nasal sac, 225

Differential hemocyte count (DHC), in invertebrates, 247, 250–251

Digestive tract, cytologic diagnosis of. See *Gastrointestinal tract lesions; Oral cavity lesions.*

Dolphins, cytologic diagnosis in. See *Cetaceans.*
  respiratory tract specimen collection in, 87–88
  urine sample collection in, 114–115, 117
  vaginal sample collection in, 121–122

Dugongs, cytologic diagnosis in. See *Sirenians.*

## E

*Edwardsiella tarda,* cytologic diagnosis of, in marine mammals, 81–82

Eels, cytologic diagnosis in, of coelomic fluid accumulation, 210
  of xanthomas, 223

Effusions, cytologic diagnosis of, 19–22
  abdominal, in ferrets, 74–75
  body cavity. See *Body cavity effusions.*
  chylous, 22
  coelomic, in amphibians, 199–200
    in fish, 210–211
    in reptiles, 170
  examination components for, 19
  exudative, 21
    in reptiles, septic vs. nonseptic, 170, 172–173

Effusions (*continued*)
  hemorrhagic, 21–22. See also *Hemorrhagic effusions/lesions*.
  malignant (neoplastic), 22
  mesothelial cell proliferation and, 20–21
  pleural, in ferrets, 74
  pseudochylous, 22
  transudative, 19–20
    in reptiles, 170–171
    modified, 170–172

Egg yolk coelomitis, cytologic diagnosis of, in avians, 146–147

Environmental contamination, of cytology specimens. See *Artifacts*.

Eosinophils, cytologic diagnosis of, in amphibians, 188
  in avians, 132–133
  in ferrets, 62
  in hemic tissue, 7–8
  in inflammation exudates, 15
  in marine mammals, 89, 92, 108–109
  in reptiles, 160

*Epistylis* sp., cytologic diagnosis of, in fish, 219–220

Epithelial carcinomas, cytologic diagnosis of, 17
  in ferrets, 64
  in reptiles, 166–167, 182

Epithelial cells, in marine mammals, conrnified vs. noncornified, 122–123
  parabasal, 105–107, 111, 122
  squamous. See *Squamous epithelial cells*.
  transitional, 117

Epithelial-glandular tissue, in cytologic diagnosis, 3

Erythrocytes, cytologic diagnosis of, in amphibians, 187–188
    inclusions and hemoparasites, 189–190
  in hemic tissue, 4–5
  in marine mammals, 91, 95, 102–103, 110, 112, 117, 122
  in reptiles, 159–161, 163, 168–169

Erythrophagocytosis, cytologic diagnosis of, in avian spleen lesions, 151
  in hemorrhagic effusions, 21–22

Esophageal lesions, cytologic diagnosis of, in avians, 143–146

Estrus cycle, in marine mammals, cytology applications of, 122–123

Extramedullary hematopoiesis, cytologic diagnosis of, in ferrets, 76–77
  in hedgehogs, 57–58

Exudative effusions, cytologic diagnosis of, 21
  in reptiles, septic vs. nonseptic, 170, 172–173

## F

Facial abscesses, cytologic diagnosis of, in prairie dogs, 48

Fatty hepatic degeneration, cytologic diagnosis of, in fish, 225–226

Feather follicles, cytologic diagnosis of, in avians, 136–140

Feathers, emerging, cytologic diagnosis of, in avians, 136–140

Feces, cytologic diagnosis of, in amphibians, 200
  in fish, 211–213
  in marine mammals, 106–113
    abnormal findings, 109–113
    normal findings, 90
    sample collection for, 106–109
  in reptiles, 185

Ferrets, cytologic diagnosis in, **61–78**
  literature review of, 61
  of infection, 62–63, 70–71, 75
  of inflammation, 62, 69–70
  of neoplasia, 63–65
  of specific anatomic sites, 65–78
    body cavity effusions, 74–75
    liver, 77–78
    lymph nodes, 72–74
    respiratory system, 75–76
    skin masses, 65–72
    spleen, 76–77
  specimen techniques for, 61

Fibroblasts, in inflammatory lesions, in avians, 133

Fibroma, cytologic diagnosis of, in avians, 138
  in ferrets, 69–70
  in fish, 221–222

Fibrosarcoma, cytologic diagnosis of, in avians, 138
    problem areas with, 153
  in ferrets, 69–70

in fish, 222
in hedgehogs, 54–55

Fibrosis, nodular dermal, in rabbits, 32–33

Fin lesions, cytologic diagnosis of, in fish, 213–214

Fine-needle aspirates (FNAs), for cytologic diagnosis, in amphibians, 191
in hedgehogs, 57, 59
in invertebrates, 236
in marine mammals, 127
in reptiles, 156–157

Fish, cytologic diagnosis in, **207–234**
benefits of, 207, 230
gill snip for, 208–209
indications for, 207
of hemorrhagic lesions, 215, 217
of infections, 210–221, 225–230
of inflammatory, 210, 212, 217, 229
of mixed cell populations, 210
of specific anatomic sites, 210–230
coelomic cavity, 210–211
digestive tract, 211–213
integument, 213–223
kidney, 223–225
liver, 225–226
reproductive tract, 227
respiratory tract, 227–228
swim bladder, 229–230
thyroid gland, 230
sample collection techniques for, 207–208
wet mount preparation for, 207–210
gills in, functional anatomy of, 227–228
kidney in, functional anatomy of, 223–224
liver in, functional anatomy of, 225

Fixative solution, for hemolymph smear preparation, 239, 250

*Flavobacterium columnare*, cytologic diagnosis of, in fish, 215–216

Flies, hemocyte identification in, 250
hemolymph collection in, 237

Fluid cytology. See *Effusions; Synovial fluid*.

Flukes, cytologic diagnosis of, in invertebrates, 244

"Follow the leader" theory, of mass standing events, 96

Foreign material, in cytology specimens. See *Artifacts*.

Formalin solution 10%, for hemolymph smear preparation, 239

Fungal infections, cytologic diagnosis of, in amphibians, 193–198, 202
in avians, 135–137, 141
gastrointestinal, 143–146
respiratory, 141–143
in ferrets, 63–64, 71–72
in fish, 216–217, 224, 227, 229
in invertebrates, 245
in marine mammals, 86
gastrointestinal, 102, 104–105, 109, 115
opportunistic, 92–93, 96–97, 118
in reptiles, 161–163

*Fusarium* sp., cytologic diagnosis of, in fish, 216–217

## G

Gallbladder lesions, cytologic diagnosis of, in fish, 226

Gastric lavage, for cytology samples, in marine mammals, 97–101
in reptiles, 181–185

Gastrointestinal tract lesions, cytologic diagnosis of, in avians, 143–146
in fish, 209, 211–213
in marine mammals, 97–106
abnormal findings, 101–106
normal findings, 90
sample collection for, 97–101
in reptiles, 181–185
cloacal swabs for, 181, 184–185
gastric lavage for, 181–185
oral swabs for, 181–183

Gerbils, mongolian, cytologic diagnosis in, 46

Gill snip, for cytologic diagnosis, in fish, 208–209, 228

Gomori's methenamine silver stain (GMS), in cytologic diagnosis, 131, 240

Gonadal tumors, cytologic diagnosis of, in fish, 227

Gout. See *Uric acid crystals*.

Gram stain, in cytologic diagnosis, 131, 135

Granulocytes, cytologic diagnosis of, in amphibians, 188–189

Granulocytes (*continued*)
 in avians, 133
 in inflammation exudates, 11–12
 in invertebrates, 248–251

Granuloma, cytologic diagnosis of, in amphibians, disseminated fungal, 196–198
 in fish, 224–226
 in marine mammals, 83
 in reptiles, 161, 163

Guinea pigs, cytologic diagnosis in, **25–49**
 of neoplasia, 35–39
  adipocyte tumors, 35–36
  bronchogenic papillary adenoma, 37
  lymphoma, 37
  mammary tumors, 36–37
  thyroid tumors, 37–38
  transitional cell carcinoma, 38–39
  trichoepithelioma, 36
  uterine leiomyoma, 37–38
 of nonneoplastic conditions, 39–40
  cervical lymphadenitis, 39
  hepatic lipidosis, 40
  Kurloff's bodies, 40
  ovarian cysts, 40
 rabbits vs., 25–35
 rodents vs., 41–48

*Gyrodactylus* sp., cytologic diagnosis of, in fish, 220–221

# H

Hamsters, cytologic diagnosis in, lymphomas, 44–45
 mammary gland adenocarcinoma, 46
 melanoma, 45
 miscellaneous tumors, 46
 papilloma, 46
 plasma cell malignancies, 44–45
 trichoepithelioma, 46

Health precautions, for working with marine mammals, 82

Hedgehogs, cytologic diagnosis in, **51–59**
 of neoplasia, 51–56
  adrenocortical carcinomas, 54–55
  fibrosarcomas, 54–55
  hemangiosarcomas, 56
  leukemia, 56
  lymphosarcoma, 53
  mammary gland adenocarcinoma, 52–53
  mast cell tumor, 53–54
  oral squamous cell carcinoma, 52
  prevalence of, 51–52
  reproductive organs, 56
  soft tissue spindle cell sarcomas, 54
  thyroid carcinomas, 55
 of nonneoplastic conditions, 56–59
  bacterial infections, 56–57
  cardiomyopathy, 58–59
  extramedullary hematopoiesis, 57–58
  hepatic lipidosis, 57, 59

Hemangioma, cytologic diagnosis of, in avians, 138
 in ferrets, 69

Hemangiosarcoma, cytologic diagnosis of, in avians, 138–139
 in ferrets, 69
 in hedgehogs, 56

Hematopoietic tumors, cytologic diagnosis of, in ferrets, extramedullary, 76–77
 in hedgehogs, extramedullary, 57–58
 in rats, 42

Hemic tissue, in cytologic diagnosis, 3–9
 basophils as, 8–9
 eosinophils as, 7–8
 erythrocytes as, 4–5
 heterophils as, 6–7
 importance of, 3–4
 leukocytes as, 4–5
 lymphocytes as, 5
 monocytes as, 5
 neutrophils as, 5–6
 platelets as, 9
 thrombocytes as, 9

Hemocyte count, total vs. differential, 247, 250–251

Hemocyte evaluation, in invertebrates, 248–251
 count examples, 247, 250–251
 cytologic diagnosis in, staining procedures for, 240
 morphologic alterations, 250
 normal identifications, 248–250
 smear preparation, 236–239

Hemocytometer, 250

Hemolymph evaluation, in invertebrates, 247–251
 clotting time, 247
 collection and smear preparation, 236–238

color, 247–248
composition, 236–237
hemocyte evaluation for, 248–251
Hemoparasites, cytologic diagnosis of, in amphibians, 190
in avians, 142–143
Hemorrhagic effusions/lesions, cytologic diagnosis of, 21–22
in avians, 138–139
in ferrets, 69
in fish, 215, 217
in hedgehogs, 56
in marine mammals, 91, 95, 102–103
in reptiles, 159–160, 162–163, 170, 173–175
Hemosiderin crystals, cytologic diagnosis of, in avians, 148–149, 151
in ferrets, 72–73
in hemorrhagic effusions, 22
Hepatic lipidosis, cytologic diagnosis of, in avians, 148–149
in guinea pigs, 40
in hedgehogs, 57, 59
in prairie dogs, 48
Hepatocellular carcinoma, cytologic diagnosis of, in prairie dogs, 47
Herpesvirus infection, cytologic diagnosis of, in fish, 213–214
in marine mammals, 86
Heterophils, cytologic diagnosis of, in hemic tissue, 6–7
in inflammation exudates, 10–11
classification of, 11–12
in avians, 132
in reptiles, 159–160, 177
mixed, 159–161, 168–169
process of, 12
Histiocytic inflammation exudates, cytologic diagnosis of, 13–14
Hormonal blood assays, in marine mammals, cytology applications of, 122–123
Hyperplasia, cytologic diagnosis of, in amphibians, chromatophore, 198–199
in fish, of thyroid, 230
tissue, 15–16

*Ichthyobodo necator,* cytologic diagnosis of, in fish, 219

*Ichthyophthirius multifiliis,* cytologic diagnosis of, in fish, 218–219

## I

Impression smears, for cytologic diagnosis, in amphibians, 190, 192–194, 201–202
in fish, 208–209, 226
in marine mammals, 127
in rabbits, 32–33
in reptiles, 157–158
Infections, cytologic diagnosis of. See also *specific classification, e.g.,* Bacterial infections.
in amphibians, 189–203
in avians, 134–137, 140–141, 147, 149
gastrointestinal, 143–146
respiratory, 141–143
in ferrets, 62–63, 70–71, 75
in fish, 210–221, 225–230
in inflammatory exudates, 12
in invertebrates, 240–241
example agents causing, 242–245
in marine mammals, 81–86
in reptiles, 161–163, 183–185
Inflammatory exudates/lesions, cytologic diagnosis of, 3, 10–15
chemotactic factors of, 10
eosinophilic, 15
granulocyte degeneration and, 11–12
heterophilic, 10–11
classification of, 11–12
process of, 12
histiocytic, 13–14
in avians, 132–134, 140
in ferrets, 62, 69–70
in fish, 210, 212, 217, 229
in invertebrates, 240–241
in marine mammals, 83–86, 90–91, 102–103, 110–113, 117
in reptiles, 159–161, 183–185
lymphocytic, 12–14
macrophagic, 12–14
melanomacrophages and, 14
mixed cell, 12–13
neutrophilic, 10–11
classification of, 11–12
process of, 12
nuclear pyknosis and, 12
phagocytic, 10–11, 14
plasmacytic, 12, 14–15
process stages of, 10
purulent characteristics in, 160–161
septic, 12

Infraorbital sinus lesions, cytologic diagnosis of, in avians, 141–143

Insects, hemocyte identification in, 249–250
  hemolymph collection in, 237
  infectious disease cytology in, 245
Integument, lesions of. See *Skin lesions*.
  of amphibians, chromatophore
      hyperplasia of, 198–199
    clinical sample techniques for,
        191–192
    functions of, 190–191
    infectious diseases of, 192–198
Interstitial cell tumor, cytologic diagnosis
    of, in rats, 42
Invertebrates, cytologic diagnosis in,
    **235–254**
      hemolymph evaluation and,
          247–251
      of infectious diseases, 240–241
        example agents causing,
            242–245
      of inflammatory lesions, 240–241
      of neoplasia, 241, 246–247
      of organs, 240
      reportable diseases, 241–245
      sample collection for, 236–238
      smear preparation for, 238–239
      species overview, 235–236,
          251–252
      staining procedures for, 240
    hemocyte evaluation in, 248–251
      collection and smear preparation,
          236–239
      count examples, 247, 250–251
      morphologic alterations, 250
      normal identifications, 248–250
    hemolymph evaluation in, 247–251
      clotting time, 247
      color, 247–248
      composition, 236–237
      hemocyte evaluation for, 248–251
Iridophores, cytologic diagnosis of, in
    amphibians, 198–199
Iridoviruses, cytologic diagnosis of, in
    amphibians, 189–190
  in fish, 214

**K**

Karyolysis, 11, 161
  cytologic diagnosis of, in marine
      mammals, 83, 85
Karyomegaly, cytologic diagnosis of, in
    avian infections, 150–151
  in ferret infections, 76–77
Karyorrhexis, 11
  cytologic diagnosis of, in marine
      mammals, 83, 85

Kidney lesions, cytologic diagnosis of, in
    avians, 140
  in fish, 209–210, 223–225
Koi, cytologic diagnosis in, of external
    parasites, 221
  of fungal infections,
      216–217
  of papillomas, 213–214
  of tumors, 222
Kupffer cells, liver inflammation and, in
    avians, 147–150
Kurloff's bodies, cytologic diagnosis of, in
    guinea pigs, 40

**L**

Larynx, of cetaceans, diagram of, 100
      uniqueness of, 98–99
Leiomyoma, cytologic diagnosis of, in
    ferrets, 69
  in guinea pigs, uterine, 37–38
Leiomyosarcoma, cytologic diagnosis of, in
    ferrets, 69
*Lernaea* sp., cytologic diagnosis of,
    in fish, 221
Lesions, cytologic diagnosis of. See *specific
    anatomy, e.g.,* Gastrointestinal tract
    lesions.
Leukemia, cytologic diagnosis of, in
    hedgehogs, 56
Leukocytes, cytologic diagnosis of, in
    amphibians, 188–189
      in hemic tissue, 4–5
      in marine mammals, 92–93,
          102, 108, 110, 113, 117
Lice, cytologic diagnosis of, in fish,
    221
Lipid-containing masses, cytologic
    diagnosis of, in avians,
    137–138
Lipid droplets, in marine mammal milk,
    125–126
Lipidosis, hepatic, cytologic diagnosis of, in
    avians, 148–149
      in guinea pigs, 40
      in hedgehogs, 57, 59
      in prairie dogs, 48
Lipoma, cytologic diagnosis of, in avians,
    137–138
      in rabbits, 29–30
      in rats, 41

Liver lesions, cytologic diagnosis of, benign. See *Hepatic lipidosis.*
    malignant. See *Hepatocellular carcinoma.*
        in avians, 147–151
        in ferrets, 77–78
        in fish, 209–210, 225–226

Lobster, hemolymph collection in, 237–238
    infectious disease cytology in, 243

Lumpy jaw, cytologic diagnosis of, in ferrets, 70

Lung lesions, cytologic diagnosis of, in avians, 143
    in ferrets, 76

Lung washes, cytologic diagnosis of, in reptiles, 178–180

Lymph sac fluid, cytologic diagnosis of, in amphibians, subcutaneous, 199–200

Lymphadenitis, cervical, cytologic diagnosis of, in guinea pigs, 39

Lymphadenopathy, cytologic diagnosis of, in ferrets, 72–74

Lymphocystis, cytologic diagnosis of, in fish, 214–215

Lymphocytes, cytologic diagnosis of, in hemic tissue, 5
    in inflammation exudates, 12–14
        in avians, 133
        in marine mammals, 108, 125–127
        in reptiles, 160

Lymphomas, cytologic diagnosis of, in avians, 137, 139
        problem areas with, 153
    in ferrets, malignant, 68–69
    in fish, malignant, 223
    in guinea pigs, 37
    in hamsters, 44–45
    in hedgehogs, malignant, 53
    in rabbits, 27–28
    in rats, 42
    in reptiles, 167, 175, 183

Lymphosarcoma (LSA), cytologic diagnosis of, in ferrets, 68–69, 73, 76–78
    in hedgehogs, 53

## M

Macrophages, in inflammation exudates, cytologic diagnosis of, 12–14
    in avian liver, 147–150
    in ferrets, 62–63, 73, 77
    in marine mammals, 95, 107, 127
    in reptiles, 160, 162–163, 175

Magnification, scanning, in cytologic diagnosis, 2
    for fish, 208
    for invertebrates, 240

Malignant effusions, cytologic diagnosis of, 22

Malignant neoplasia, cytologic diagnosis of, 16–19. See also *specific pathology or tumor, e.g., Lymphoma.*
    cytoplasmic cellular features of, 16–17
    diagnostic categories for, 16
    effusions, 22
    epithelial cell origins, 17
    general cellular features of, 16, 136
    in amphibians, 203–204
    in avians, 135–136
    in ferrets, 63–65
    in fish, 223
    in reptiles, 163–168
    mesenchymal origins, 17–18
    nuclear features of, 16–17
    nuclear features seen in, 164–165
    round cell origins, 18–19
    species indications for. See *specific species.*
    structural features of, 16–19

Mammary gland tumors, cytologic diagnosis of, in guinea pigs, 36–37
    in hamsters, 46
    in hedgehogs, 52–53
    in marine mammals, 124–127
    in rabbits, 26–28
    in rats, 41

Manatees, cytologic diagnosis in. See *Sirenians.*
    gastric specimen sampling in, 105–106

Marine mammals, cytologic diagnosis in, **79–130**. See also *Cetaceans; Sirenians.*
    gastric specimen sampling in, 97–101, 105–106
    respiratory tract specimen collection in, 86–87
    urine sample collection in, 114–115, 117
    vaginal sample collection in, 121–122

*Marteilia refringens,* cytologic diagnosis of, in invertebrates, 241–242, 246

Mast cell tumor, cytologic diagnosis of, in ferrets, 65–66
    in hedgehogs, 53–54

Mast cells, in inflammatory lesions, in avians, 133

Melanocytes, in body cavity effusions, of reptiles, 171

Melanoma, cytologic diagnosis of, in fish, 222–223
    in hamsters, 45

Melanomacrophage centers (MMCs), in fish, 225

Melanomacrophages, cytologic diagnosis of, in amphibians, 201–202
    in inflammation exudates, 14

Melanophores, cytologic diagnosis of, in amphibians, 198

Mesenchymal neoplasms, cytologic diagnosis of, 17–18
    in ferrets, 63–64, 69–70
    in fish, 222
    in reptiles, 166–169

Mesomycetozoea, cytologic diagnosis of, in amphibians, 197

Mesothelial cells, proliferation in effusions, 20–21
    in ferrets, 75
    in reptiles, 169–171

Mesothelioma, cytologic diagnosis of, in ferrets, 75
    in reptiles, 175

Metazoan parasites, cytologic diagnosis of, in fish, 211–212

Methenamine silver stain, Gomori's, in cytologic diagnosis, 131, 240

Methylene blue, in cytologic diagnosis, 3
    for reptiles, 159

Mice, cytologic diagnosis in, bacterial infections, 44
    neoplasia, 43

*Microsporum canis,* cytologic diagnosis of, in ferrets, 71

Milk specimens, cytologic diagnosis of, in marine mammals, 123–127
    abnormal findings, 125–127
    normal findings, 90
    sample collection for, 123–125

Mineral deposits, in synovial fluid, of avians, 147–148
    of reptiles, 177–179

Mite infestations, cytologic diagnosis of, in avians, 136–137, 142

Mitotic figures, in cytologic diagnosis, 2
    in reptiles, 164–165, 167

Mixed cell lesions, cytologic diagnosis of, in fish, 210
    in reptiles, 159–161, 168–169
    inflammatory, 12–13

Molding, nuclear, in malignant neoplasms, 164–165

Molds, oomycetes water, in amphibians, 197

Mollusks, bivalve, hemocyte identification in, 249, 251
    hemolymph collection in, 237
    infectious disease cytology in, 242–243

Monocytes, in hemic tissue, cytologic diagnosis of, 5

Morbillivirus infection, cytologic diagnosis of, in marine mammals, 86

Mosquitoes, hemolymph collection in, 237

Mucin, in synovial fluid, of reptiles, 176

Mucocele, salivary gland, in ferrets, 72–73

Mucopolysaccharides, in cytologic diagnosis, 3

*Mucor amphibiorum,* cytologic diagnosis of, in amphibians, 196–197

Multinucleation, in malignant neoplasms, 164–166

Mycobacteriosis, cytologic diagnosis of, in amphibians, 192–194, 202
    in ferrets, 70–71
    in fish, 216–217, 225, 229
    in marine mammals, 82

Myelolipoma, cytologic diagnosis of, in avians, 137, 139–140

Myxoma, cytologic diagnosis of, in ferrets, 69

Myxosarcoma, cytologic diagnosis of, in ferrets, 69

Myxosporean species, cytologic diagnosis of, in fish, 226–227, 229

Myxozoa, cytologic diagnosis of, in
 amphibians, 202–203
 in marine mammals, 106, 110

# N

Nasal sac, cytologic diagnosis of, in marine
 mammals, 89–95
 sampling technique for,
 87–88

Nasal sinus lesions, cytologic diagnosis of,
 in avians, 141–143

Nasal swabs, cytologic diagnosis of, in
 marine mammals, 86–93
 in reptiles, 178–179

*Nasitrema attenuata,* cytologic diagnosis of,
 in marine mammals, 93, 96, 104–105

Nematodes, cytologic diagnosis of, in
 amphibians, 190, 198, 203
 in fish, 211–212, 229
 in marine mammals,
 gastrointestinal, 104

Neoplasia, cytologic diagnosis of, benign,
 15–16. See also *specific pathology or
 tumor, e.g.,* Papilloma.
 effusions and, 22, 170, 175
 in avians, 135–136
 problem areas with,
 152–153
 in ferrets, 63–65
 metastatic, 74
 in fish, 221–223
 in guinea pigs, 35–39
 in hedgehogs, 51–56
 in invertebrates, 241, 246–247
 in mongolian gerbils, 46
 in prairie dogs, 47
 in rabbits, 25–31
 in rats, 41–42
 in reptiles, 159–160, 163–168
 malignant, 16–19. See also
  *specific pathology or tumor,
  e.g.,* Lymphoma.
  cytoplasmic cellular features
   of, 16–17
  diagnostic categories for, 16
  epithelial cell origins, 17
  general cellular features
   of, 16
  mesenchymal origins,
   17–18
  nuclear features of, 16–17
  round cell origins, 18–19
  structural features of,
   16–19

Nervous tissue, in cytologic diagnosis, 3

Neuroendocrine neoplasms, cytologic
 diagnosis of, in reptiles, 167–168

Neurofibromas, cytologic diagnosis of, in
 fish, 221–222

Neutrophils, cytologic diagnosis of, in
 amphibians, 188
 in hemic tissue, 5–6
 in inflammation exudates, 10–11
  classification of, 11–12
  process of, 12
 in marine mammals, 83, 85, 89,
  92, 95, 103, 108–110,
  122

Nodular dermal fibrosis, cytologic diagnosis
 of, in rabbits, 32–33

Noncellular characteristics, in cytologic
 diagnosis, 2–3

Nonneoplastic conditions, cytologic
 diagnosis of, in chinchillas, 48
 in guinea pigs, 39–40
 in hedgehogs, 56–59
 in mongolian gerbils, 46
 in prairie dogs, 48
 in rabbits, 31–35
 in rats, 42–43
 in sugar gliders, 48

Nuclear features, of malignant neoplasms,
 164–165

Nuclear pyknosis, in inflammation
 exudates, cytologic diagnosis of, 12

Nuclear to cytoplasmic ratio (N:C ratio), 2
 in malignant neoplasia, 17, 164–165
 of reptiles, 164

Nucleoli, in malignant neoplasms,
 164–165

# O

Odontoma, cytologic diagnosis of, in prairie
 dogs, 47

Oomycetes water molds, cytologic diagnosis
 of, in amphibians, 197

Oral cavity lesions, cytologic diagnosis of, in
 avians, 143–146
 in ferrets, 72
 in reptiles, 182–183

Oral squamous cell carcinoma, cytologic
 diagnosis of, in hedgehogs, 52

Oral swabs, for cytologic diagnosis, in
 reptiles, 181–183

Orogastric tube, for cytologic samples, in
 marine mammals, 97–101

Oropharyngeal swabs, cytologic diagnosis of, in reptiles, 179

Osteosarcoma, cytologic diagnosis of, in reptiles, 168

Ovarian cysts, cytologic diagnosis of,
  in fish, 227
  in guinea pigs, 40

Ovarian tumors, cytologic diagnosis of, in hedgehogs, 56
  in mongolian gerbils, 46

Oyster, hemocyte identification in, 249, 251
  infectious disease cytology in, 242, 246

## P

Papillary adenoma, bronchogenic, cytologic diagnosis of, in guinea pigs, 37

Papilloma, cytologic diagnosis of, in fish, 213–214
  in hamsters, 46
  in rabbits, squamous, 32–33

Papillomavirus infection, cytologic diagnosis of, in marine mammals, 86

Parabasal epithelium, cytologic diagnosis of, in marine mammals, 105–107, 111, 122

Parasite infections, cytologic diagnosis of, in amphibians, of blood, 189–190
    of skin, 198
    of visceral organs, 202–203
  in avians, 135–136, 149
    gastrointestinal, 143–146
    respiratory, 141–143
  in fish, external, 218–222, 228
    internal, 211–212, 224–225, 227
  in invertebrates, 241–242, 244
  in marine mammals, 93, 96–97, 108
    gastrointestinal, 103–105, 110, 112
  in reptiles, 185

Periodic acid-Schiff stain, in cytologic diagnosis, 131, 134
  for fish, 214–215
  for invertebrates, 240–241

Peripheral blood, cytologic diagnosis of, contamination with, 3, 21
  in amphibians, 187–190, 194

"*Perkinsus* -like" protozoa, cytologic diagnosis of, in amphibians, 202–203

pH, in marine mammals, gastric sampling of, 100–102, 105
  urine sampling of, 120

Phaeohyphomycosis, cytologic diagnosis of, in amphibians, 193–194

Phagocytes, cytologic diagnosis of, in inflammation exudates, 10–11, 14
  in reptiles, 159
  in marine mammals, 83, 85

Pharynx, of cetaceans, uniqueness of, 98–99

Pigment cell tumors, cytologic diagnosis of, in fish, 222–223
  in hamsters, 45

Plant debris, cytologic diagnosis of, in marine mammals, 97, 105–106, 115

Plasma cell malignancies, cytologic diagnosis of, in ferrets, 76
  in hamsters, 44–45

Plasma cells, in inflammation exudates, cytologic diagnosis of, 12, 14–15
  in avians, 133

*Plasmodium* sp., cytologic diagnosis of, in avians, 149–150

Platelets, in hemic tissue, cytologic diagnosis of, 9

*Pleistophora* sp., cytologic diagnosis of, in fish, 218

Pleomorphism, 164–165

Pleural effusion, cytologic diagnosis of, in ferrets, 74

Pneumonia, cytologic diagnosis of, in ferrets, 75–76
  in reptiles, 180–181

Pollen particles, cytologic diagnosis of, in marine mammals, 97, 116

Polycystic kidney disease, cytologic diagnosis of, in fish, 224–225

Polyps, cytologic diagnosis of, in rabbits, 31–32

Poxvirus infection, cytologic diagnosis of, in marine mammals, 86

Prairie dogs, cytologic diagnosis in, of neoplasia, 47
    hepatocellular carcinoma, 47
    odontoma, 47
  of nonneoplastic conditions, 48

facial abscesses, 48
hepatic lipidosis, 48
miscellaneous tumors, 48
necrotic stomatitis, 48
Prawns, infectious disease cytology in, 243
Preservation techniques, for hemolymph smear preparation, 239, 250
Proliferative lesions, of ventral abdominal marking gland, in hamsters, 46
in mongolian gerbils, 46
Protective measures, for working with marine mammals, 82
Proteins, in cytologic diagnosis, 3
Protozoa, cytologic diagnosis of, in amphibians, 190, 202–203
in avians, 142–143
in fish, 212–213, 219, 226
in invertebrates, 241–242
in marine mammals, 82, 96, 114
Pseudochylous effusions, cytologic diagnosis of, 22
Public health, cytologic diagnosis considerations of, in marine mammals, 81–82
Pyknosis, cytologic diagnosis of, in marine mammals, 83, 85
nuclear, in inflammation exudates, 12
Pyoderma, cytologic diagnosis of, in rats, 43
Pyometra, cytologic diagnosis of, in rats, 42–43

## Q

Quality impressions, in cytologic diagnosis, 2

## R

Rabbits, cytologic diagnosis in, **25–49**
guinea pigs vs., 35–40
of neoplasia, 25–31
basal cell tumors, 25–26
lipoma, 29–30
lymphoma, 27–28
mammary gland tumors, 26–28
soft tissue sarcoma, 28–29
squamous cell carcinoma, 26
testicular tumors, 30–31
thymoma, 31
trichoepitheliomas, 26–27
uterine adenocarcinoma, 29–30
of nonneoplastic conditions, 31–35
abscesses, 33–34
cutaneous treponemiasis, 33–35
nodular dermal fibrosis, 32–33
polyps, 31–32
squamous papilloma, 32–33
rodents vs., 41–48
Rats, cytologic diagnosis in, of neoplasia, 41–42
hematopoietic tumors, 42
interstitial cell tumor, 42
lipoma, 41
lymphoma, 42
mammary gland tumors, 41
Zymbal's gland tumor, 41
of nonneoplastic conditions, 42–43
bronchopneumonia, 42
pyoderma, 43
pyometra, 42–43
vaginitis, 42–43
Rectal tube, for cytologic samples, in marine mammals, 107, 110
Renal carcinoma, cytologic diagnosis of, in avians, 140
Renal tubular cells, cytologic diagnosis of, in marine mammals, 117–118
Reportable diseases, in invertebrates, cytologic diagnosis of, 241–245
Reproductive organ lesions, cytologic diagnosis of, in avians, 147
in fish, 227
in guinea pigs, 37–38, 40
in hedgehogs, 56
in rabbits, 29–30
Reptiles, cytologic diagnosis in, **155–186**
critical rule for, 155–156
of body cavity effusions, 169–175
approach to, 169–170
coelomic classifications of, 170
exudates, septic vs. nonseptic, 170, 172–173
hemorrhagic, 170, 173–175
melanocytes in, 171
mesothelial cells in, 169–171

Reptiles (*continued*)
    modified transudates, 170–172
    neoplastic, 170, 175
    transudates, 170–171
  of cystic lesions, 160–162
  of fluids, body cavity effusions, 169–175
    synovial fluid, 175–179
  of gastrointestinal tract, 181–185
    cloacal swabs for, 181, 184–185
    gastric lavage for, 181–185
    oral swabs for, 181–183
  of hemorrhagic lesions, 159–160, 162–163
  of inflammatory lesions, 160–161
  of mixed cell population, 159–161, 168–169
  of neoplastic lesions, 159–160, 163–168
    cytoplasmic features in, 166
    gross characteristics in, 163–164
    malignancy criteria, 166
    nuclear features in, 164–165
    specific tumors, 166–168
  of respiratory tract, 178–181
    microorganisms in, 180–181
    normal components in, 179–180
    specimen collection for, 178–179
  sample collection for, 156–158
  slide preparation for, 158–159
  species overview, 155, 185
  tissue aspirate categories for, 159–160

Respiratory tract lesions, cytologic diagnosis of, in avians, 141–143
  lower, 143
  upper, 141–143
  in ferrets, 75–76
  in fish, 227–228
  in marine mammals, 86–97
    abnormal findings, 89, 91–97
    normal findings, 90
    sample collection for, 86–89
  in reptiles, 178–181
    microorganisms in, 180–181
    normal components in, 179–180
    specimen collection for, 178–179

Restraint, for cytology sample collection, in marine mammals, 82–83

Rodents, cytologic diagnosis in, **25–49**
  chinchillas, 48
  guinea pigs vs., 35–40
  hamsters, 44–46
  mice, 43–44
  mongolian gerbils, 46
  prairie dogs, 46–48
  rabbits vs., 25–35
  rats, 41–43
  sugar gliders, 48

Romanowsky-type stains, in cytologic diagnosis, 1
  for avians, 131, 134–135
  for ferrets, 62
  for invertebrates, 6–7
  for reptiles, 158–159

Round cell tumors, cytologic diagnosis of, 18–19
  in amphibians, 203–204
  in ferrets, 64
  in fish, 223
  in reptiles, 167, 182–183

## S

Salivary gland mucocele, cytologic diagnosis of, in ferrets, 72–73

Salt crystals, in cytologic samples, of marine mammals, 99

*Sarcocystosis falcatula,* cytologic diagnosis of, in avians, 143–144

Sarcomas, cytologic diagnosis of. See also *specific pathology, e.g.,* Soft tissue sarcoma.
  in reptiles, 182

Schwannomas, cytologic diagnosis of, in fish, 222

Scorpion, hemolymph collection in, 238

Sebaceous cysts, cytologic diagnosis of, in reptiles, 161–162

Sebaceous gland tumors, cytologic diagnosis of, in ferrets, 66–67

Secretory cells, in cytologic diagnosis, 2–3

Seminoma, cytologic diagnosis of, in fish, 227

Septic inflammation exudates, cytologic diagnosis of, 12. See also *Infections.*
  in reptiles, 161, 170, 172
    gastrointestinal disease and, 183–185
    respiratory disease and, 180

Septicemia, bacterial hemorrhagic, cytologic diagnosis of, in fish, 215

Shellfish, bivalve, hemolymph collection in, 237
    infectious disease cytology in, 242–243

Sialocele, cytologic diagnosis of, in ferrets, 72–73

Silkworm, infectious disease cytology in, 245

Sinus lesions, cytologic diagnosis of, in avians, 141–143

Sirenians, cytologic diagnosis in, **79–130**
    literature review of, 81
    of feces, 106–109, 111–113
    of gastrointestinal tract, 97–101, 105–106
    of infections, 81–86
    of inflammatory lesions, 83–86, 90–91, 102–103, 111–113
    of milk, 123–127
    of miscellaneous specimens, 127
    of respiratory tract, 86–93
    of urine, 113–120
    pathologic significance evaluation, 83–86
    public health considerations, 81–82
    restraint for sample collection, 82–83
    species overview, 79–81, 127

Skin lesions, cytologic diagnosis of, in amphibians, 190–199
    in avians, 136–140
    in ferrets, 65–72
    in fish, infectious, 213–221
        neoplastic, 221–223
    in marine mammals, 127

Skin scrapings, for cytologic diagnosis, in amphibians, 191, 194–199
    in fish, 207
    in invertebrates, 236
    in rabbits, 35
    in reptiles, 156–157

Smear preparations, for cytologic diagnosis. See *specific species or sample.*
    impression. See *Impression smears.*

Snails, hemolymph collection in, 237–238

Soft tissue sarcoma, cytologic diagnosis of, in hedgehogs, spindle cell, 54
    in rabbits, 28–29

Specimens, for cytology, collection techniques for. See *specific species.*
    contamination of. See *Artifacts.*

Spermatozoa, cytologic diagnosis of, in marine mammals, 118–119

Spider, hemocyte identification in, 249–250
    hemolymph collection in, 237
    infectious disease cytology in, 245

Spindle cell sarcomas, cytologic diagnosis of, in hedgehogs, 54

Spindle cell tumors, cytologic diagnosis of, in amphibians, 203
    in avians, 137–139
    in ferrets, 64, 69–70
    in fish, 221–222
    in hamsters, 46

Spleen lesions, cytologic diagnosis of, in avians, 147, 151
    in ferrets, 76–77

Squamous cell carcinoma, cytologic diagnosis of, in avians, 137, 140–141
    in ferrets, 69
    in hedgehogs, oral, 52
    in rabbits, 26

Squamous epithelial cells, in marine mammals, artifacts vs., 97–98, 103, 105–106, 112–113, 115, 125
    gastrointestinal, 102, 104–107, 110–112
    respiratory, 89, 91, 94, 96
    urinary, 117–118

Squamous papilloma, cytologic diagnosis of, in rabbits, 32–33

Squash preparation, for cytologic diagnosis, in fish, 209–212, 224–225, 227

Stains, in cytologic diagnosis, 1–2. See also *specific stain.*
    for amphibians, 193
    for avians, 131
    for invertebrates, 240, 250–251
    for reptiles, 158–159

Stomatitis, necrotic, cytologic diagnosis of, in prairie dogs, 48

Subcutis lesions, cytologic diagnosis of, in avians, 136–140

Sugar gliders, cytologic diagnosis in, 48

Swab specimens, for cytologic diagnosis, in
  avians, 142, 145–146
    in invertebrates, 236
    in marine mammals, 86–93
    in rabbits, 43
    in reptiles, 158–159, 178–179,
      181–185
Swim bladder, cytologic diagnosis of, in
  fish, 229–230
Synovial cell sarcoma, cytologic diagnosis
  of, in ferrets, 69
Synovial fluid, cytologic diagnosis of, in
  avians, 147–148
    in reptiles, 175–179
Syphilis, cytologic diagnosis of, in rabbits,
  33–35

**T**

Tarantula, hemocyte identification in,
  249–250
Testicular tumors, cytologic diagnosis of, in
  rabbits, 30–31
Thrombocytes, in hemic tissue, cytologic
  diagnosis of, 9
Thymoma, cytologic diagnosis of, in
  rabbits, 31
Thyroid tumors, cytologic diagnosis of, in
  fish, 230
    in guinea pigs, 37–38
    in hedgehogs, 55
Tissue classifications, in cytologic
  diagnosis, 3
Tissue hyperplasia, cytologic diagnosis of,
  15–16
Total hemocyte count (THC), in
  invertebrates, 247, 250–251
Tracheal lesions, cytologic diagnosis of, in
  avians, 141–143
    in ferrets, 75–76
Transitional cell carcinoma, cytologic
  diagnosis of, in guinea pigs, 38–39
Transitional epithelial cells, in marine
  mammals, 117
Transtracheal wash, cytologic diagnosis of,
  in reptiles, 178–181
Transudative effusions, cytologic diagnosis
  of, 19–20
    in reptiles, 170–171
      modified, 170–172

Trematodes, cytologic diagnosis of, in fish,
  212, 220–221
    in marine mammals, 93, 96
      gastrointestinal,
        104–105
Treponemiasis, cutaneous, cytologic
  diagnosis of, in rabbits, 33–35
Trichodinid sp., cytologic diagnosis of, in
  fish, 219–220
Trichoepithelioma, cytologic diagnosis of,
  in guinea pigs, 36
    in hamsters, 46
    in rabbits, 26–27
Tumors. See *Neoplasia; specific anatomy or
  pathology.*

**U**

Ulcer disease, cytologic diagnosis of, in fish,
  215, 217
    in marine mammals, 103
Uric acid crystals, in synovial fluid, of
  avians, 147–148
    of reptiles, 177–178
Urinary casts, cytologic diagnosis of, in
  marine mammals, 118–120
Urinary catheter, for cytologic samples, in
  marine mammals, 114–115, 117
Urine, cytologic diagnosis of, in marine
  mammals, 113–120
    abnormal findings,
      117–120
    normal findings, 90
    sample collection for,
      113–117
Uroliths, cytologic diagnosis of, in marine
  mammals, urinary, 118–120
Uterine adenocarcinoma, cytologic
  diagnosis of, in rabbits, 29–30
Uterine carcinoma, cytologic diagnosis of,
  in hedgehogs, 56
Uterine leiomyoma, cytologic diagnosis of,
  in guinea pigs, 37–38

**V**

Vaginal specimens, cytologic diagnosis of,
  in marine mammals, 121–123
    abnormal findings,
      121–123
    normal findings, 90
    sample collection for,
      121

Vaginitis, cytologic diagnosis of, in rats, 42–43

Vent lesions, cytologic diagnosis of, in avians, 145–146

Ventral abdominal marking gland, proliferative lesions of, in hamsters, 46
    in mongolian gerbils, 46

*Vibrio* sp., cytologic diagnosis of, in marine mammals, 81–82

Viral infections, cytologic diagnosis of, in amphibians, 189–190, 240
    in avians, 135–136, 140–141, 149–150
        gastrointestinal, 143–146
        respiratory, 141–143
    in ferrets, 75
    in fish, 213–215
    in invertebrates, 241–243, 245
    in marine mammals, 86

Visceral organ lesions, cytologic diagnosis of. See also *specific organ, e.g.,* Spleen lesions.
    in amphibians, 200–203
    in fish, 209–210, 214

## W

Wet mount preparations, for cytologic diagnosis, in amphibians, 191

    in fish, 207–210
    in marine mammals, 117

Windrowing, 176–177

Wright's stain, in cytologic diagnosis, 1–3, 240
    of marine mammals, 88, 101, 109, 117, 121, 125

## X

Xanthoma, cytologic diagnosis of, in avians, 137–138
    in fish, 223

Xanthophores, cytologic diagnosis of, in amphibians, 198

## Y

Yeast infections, cytologic diagnosis of, in avians, 144
    in marine mammals, 85–86, 110

## Z

Zoonotic infection, cytologic diagnosis of, in marine mammals, 81

Zymbal's gland tumor, cytologic diagnosis of, in rats, 41

# Moving?

## Make sure your subscription moves with you!

To notify us of your new address, find your **Clinics Account Number** (located on your mailing label above your name), and contact customer service at:

E-mail: elspcs@elsevier.com

800-654-2452 (subscribers in the U.S. & Canada)
407-345-4000 (subscribers outside of the U.S. & Canada)

Fax number: 407-363-9661

**Elsevier Periodicals Customer Service**
6277 Sea Harbor Drive
Orlando, FL 32887-4800

*To ensure uninterrupted delivery of your subscription, please notify us at least 4 weeks in advance of move.